POLITICAL THEOLOGY OF THE EARTH

INSURRECTIONS: CRITICAL STUDIES
IN RELIGION, POLITICS, AND CULTURE

INSURRECTIONS: CRITICAL STUDIES IN RELIGION, POLITICS, AND CULTURE

SLAVOJ ŽIŽEK, CLAYTON CROCKETT, CRESTON DAVIS, JEFFREY W. ROBBINS, EDITORS

The intersection of religion, politics, and culture is one of the most discussed areas in theory today. It also has the deepest and most wide-ranging impact on the world. Insurrections: Critical Studies in Religion, Politics, and Culture will bring the tools of philosophy and critical theory to the political implications of the religious turn. The series will address a range of religious traditions and political viewpoints in the United States, Europe, and other parts of the world. Without advocating any specific religious or theological stance, the series aims nonetheless to be faithful to the radical emancipatory potential of religion.

After the Death of God, John D. Caputo and Gianni Vattimo, edited by Jeffrey W. Robbins

The Politics of Postsecular Religion: Mourning Secular Futures, Ananda Abeysekara

Nietzsche and Levinas: "After the Death of a Certain God," edited by Jill Stauffer and Bettina Bergo

Strange Wonder: The Closure of Metaphysics and the Opening of Awe, Mary-Jane Rubenstein

Religion and the Specter of the West: Sikhism, India, Postcoloniality, and the Politics of Translation, Arvind Mandair

Plasticity at the Dusk of Writing: Dialectic, Destruction, Deconstruction, Catherine Malabou

Anatheism: Returning to God After God, Richard Kearney

Rage and Time: A Psychopolitical Investigation, Peter Sloterdijk

Radical Political Theology: Religion and Politics After Liberalism, Clayton Crockett

For a complete list of books in this series, see page 231

POLITICAL THEOLOGY OF THE EARTH

OUR PLANETARY EMERGENCY AND THE STRUGGLE FOR A NEW PUBLIC

CATHERINE KELLER

Columbia University Press *New York*

Columbia University Press
Publishers Since 1893
New York   Chichester, West Sussex
cup.columbia.edu
Copyright © 2018 Columbia University Press
All rights reserved

Library of Congress Cataloging-in-Publication Data
Names: Keller, Catherine, 1953– author.
Title: Political theology of the earth : our planetary emergency and the struggle for a new public / Catherine Keller.
Description: New York : Columbia University Press, [2018] |
Series: Insurrections : critical studies in religion, politics, and culture |
Includes bibliographical references and index.
Identifiers: LCCN 2018024685 | ISBN 9780231189903 (cloth : alk. paper) | ISBN 9780231189910 (pbk. : alk. paper) | ISBN 9780231548618 (e-book)
Subjects: LCSH: Political theology. | Public theology. | Religion and politics. | Ecology—Religious aspects. | Ecology—Political aspects.
Classification: LCC BL65.P7 K45 2018 | DDC 201/.77—dc23
LC record available at https://lccn.loc.gov/2018024685

Columbia University Press books are printed on permanent and durable acid-free paper.
Printed in the United States of America

Cover design: Alex Camlin

To my students and colleagues, at this very moment creating the world that can be.

CONTENTS

Acknowledgments ix

Beginning 1

1 Political: Sovereign Exception or Collective Inception 21

2 Earth: Climate of Closure, Matter of Disclosure 69

3 Theology: "Unknow Better Now" 105

Apophatic Afterword 159

Notes 181
Index 215

ACKNOWLEDGMENTS

In a time of heightened political disarray and lowered planetary hope, one appreciates more vividly the solidarity and the support of persons, collectives, even institutions. This is not a matter of routine interdependence, which requires an ontology but no special thanks. So let me first express my gratitude to Yale Divinity School for the invitation to do the Nathaniel W. Taylor Lectures, February 2017, which provoked an early draft of *Political Theology of the Earth*. And thanks to the American Theological Association for the Luce Grant that allowed me to finish the book in the timely manner its subject matter demanded; it will lead to *Apocalypse After All?*, the main project for the leave. And, as forever, I am indebted to my home institution Drew, the Theological School, whose dean pressed me to apply for that grant; and whose motto of courage, innovation, and rootedness actually holds true. More institutions: it was a pleasure to deliver the lectures at Mercer University under the same heading, about when I was delivering the manuscript. And I thank Columbia University Press for taking it from there, and especially the subtle Wendy Lochner, for her indelible influence at every stage. I am also indebted, once again, to Susan Pensak for her finely tuned copyediting.

Which has brought me to persons and their gifts. First of all I thank the poet Ed Roberson for his gracious permission to cite so much of his piercing poetry—the sine qua non of my Earth chapter and so of this book.

I will not thank enough the all-too-gifted Winfield Goodwin, who as my research assistant offered not just editorial support from beginning to end but contributed endless sources and clarifications. Furthermore, I was fortunate to receive a treasure trove of suggestions from the radical political theologian Clayton Crockett, all of which I greedily took. And later I received, to my delight, a full and indispensable reading from my doctoral student Anna Blaedel. Past and present doctoral students consistently inform and inspire my work; and among those who precede and inform me in the explicit engagement of political theology I will here just name Dhawn Martin, Yountae An, Karen Bray, Shelley Yael Dennis, Elijah Prewitt-Davis, and Austin Roberts.

It is impossible to enumerate here, after this short book, the cloud of friends whose conversation, intervention, interest, and writing colludes in this and any work I do.

I do though thank, for innumerable gifts, my person of persons Jason Starr.

POLITICAL THEOLOGY OF THE EARTH

BEGINNING

Once upon a time we had . . . time.
Whatever the story of our individual mortalities, there extended out from all of us, from us all together, the space of a shared time, the time of a shared space. The sharing was rent with contradiction: we reached no consensus on the layout of the future. We could ignore the space of its temporal bodies and squint away the alpha and the omega of its ages. Our calculations collided, our opposed futures warred and left hope drugged or in ruins. But still there stretched before us—if we were not fundamentalists of The End—at least a time to rebuild. There would be time enough for the space of a more marvelous togetherness: New Heaven and Earth, utopic horizon, seventh generation, endless rhythm, eternal return, r/evolutionary leap, fitful progress, sci-fi tomorrow. Or so the stories go. We had time.

And now we seem to have lost it.

Time, our time, the time of human civilization, appears to be running out. The science of climate has been unhysterically, relentlessly, increasingly signaling: not that time *will* run out but that *if* we stay on present course . . . So it had seemed, at least before the acceleration expressed in the 2016 U.S. election, we had a fighting chance of changing course within the narrow window of time that climate change allots. After the political shift, however, the window seemed to be slamming shut. Not on all of life, not on the earth, not necessarily even on our species. But on historic human civilization as it flows into its future. Yet it is precisely so-called civilization that had brought us to this moment of self-contradiction, at which point we would be too busy responding politically to immediate threats to vulnerable human populations—of black lives, of immigrant, uninsured, or sexually abused lives—to mind the matter of the earth.

How could we answer new threats by and to the political process itself and, at the same time, attend to the inhuman elements of water, atmosphere, land, warmth, to the global economics of extraction, not to mention, say, to the endangered elephants? Even if we insist—as many of us will have continued to do—that social justice and ecological viability come inextricably entangled?

Am I writing, then, to proclaim the window shut? To perform just the sort of self-fulfilling prophecy of The End that tempts the left with paralysis? Conveniently for the right, we would thus shut ourselves down. Succumbing to a reasonable hopelessness, a critically plausible nihilism, we become one—in effect—with reactionary denialism.

KAIROS AND CONTRACTION

So, no, I wouldn't bother to write, nor you, I wager, to read, if the window had closed. But it does seem to be closing. No way around it: the time is short. Really then I seem to be just quoting a text favored by the apocalyptic Christian right: "The appointed time is short" (1 Cor. 7:29). The apostle Paul, however, surely did not mean by "short" two thousand years or so. Nor can he here be read as making an appointment with ecological catastrophe (unlike his neighbor in the Second Testament, John of Patmos, who can be).

I did however make an appointment with Paul. This was odd, given that I had hardly outgrown the entrenched, twentieth-century feminist theological habit of opposing liberating/Jewish/prophetic gospel to sexist/heterosexist/supercessionist epistle. Given the apostle's indispensable appropriation by every form of Christian right-wing or mainline orthodoxy, it took the disarming enchantment with Paul of a growing assemblage of continental political thinkers—mostly not-Christian, atheist, or more or less Marxist—to cut through my encrusted suspicion.

And, lo, what did I find but that the trustworthy New Revised Standard Version's translation ("The appointed time is short") misleads on two counts. The Greek word translated "short" is far more complicated, more inviting, indeed more political: *sunestalemnos* means "gathered together," "contracted." It was not from a biblical scholar but from the political philosopher Giorgio Agamben, in his meditation on Paul, *The Time That Remains*, that I got this clue.[1] Also the "appointed" of "appointed time" translates better as "remaining." Agamben does not note that the predicate in his titular citation seems to be syntactically misplaced, that the phrase may be more accurately translated simply as "the time is contracted."

In other words, Paul is not (on this Agamben is clear) announcing some predetermined end of the time line or programming an upcoming appointment with Christ at the Second Coming. Biblical scholars argue that a significant shift in Paul's thinking has taken place since his earlier letter to the Thessalonians, where he may well have expected an imminent End.[2]

The Corinthian word for "time" is not the standard *chronos*. According to another Paul, the original meaning of *kairos*, "the right time, the time in which something can be done . . . must be contrasted with *chronos*, measured time or clock time. The former is qualitative, the latter quantitative."[3] In contrast to the chronological continuum of calculable time, Paul Tillich thus surfaced for theology, "in the context of religious socialism," the eventive moment of *kairos*.⁴ The kairos signifies a breakthrough into, not out of, concrete history. For classical rhetoric and Stoicism (in which the apostle was well versed), the temporality of kairos had named the opportune or critical moment, "a passing instant when an opening appears which must be driven through with force if success is to be achieved."⁵ It was first used of arrows but also of the weaving loom's shuttle (the speed of which I witnessed as a child in Greece). In the immediate wake of apartheid, the kairos documents of South Africa swiftly initiated, for the sake of justice rather than vengeance, a political practice of messianic eventiveness.⁶

In Agamben's analysis, "*kairos is a contracted and abridged chronos.*"⁷ This captures an intensification through which, as we shall see, he channels the political messianism of Walter Benjamin. Not far from the politics of religious socialism, Benjamin's early twentieth-century "now-time" (*Jetztzeit*) has become key to the current conversations in political theology with which this book is imbricated.⁸ The notion of abridgment can be misread as shrinkage—again, as mere lack of time. Yet for Paul "the kairos is filled full."⁹ As the biblical scholar L. L. Welborn puts it in his highly contracted Pauline commentary, "the kairos arrests and suspends chronos."¹⁰ Agamben twists the kairos differently, avoiding a dualism of secular immanence and extratemporal transcendence: "Messianic time is that part of secular time which undergoes an entirely transformative contraction."¹¹

The empty continuity of chronotime is interrupted by a messianic contraction, in a decisive "now" growing from the radical gospel teaching of the "kingdom of God"—that is, of a politico-spiritually charged transformation, immanent to the event of the kairos.¹² So then Paul's point will not have been to menace the community at Corinth with the end of the world. The letter says just three verses later: "For the present *schema* of this world is passing away . . ." *Schema* means "form," "order," "schematism"—suggestive both the theory of a worldview and of politico-economic shapes of power. What was to end is a human construction of the world, not "the world" itself. And, he adds, precisely to dispel the paralyzing affect of doom in the face of real crisis: "I want you to be free of anxieties" (1 Cor. 7:32).

I want that too.

WITHOUT EXCEPTION

In that spirit and in this time, we contemplate together *a political theology of the Earth*. In this Anthropocene moment of mounting crisis, the schema of the world—not the earth itself but the schematism of a civilization based on eleven thousand years of Holocene climate stability—does seem to be passing away. Schemes within schemes—"ancient schemes" of religion within politics now come buried within modern schemes of politics trapped in economic schemes, and wrapped within the planetary scene of climate change.[13] There, human scheming seems to be finding the limit not of its impact but of its control. And so of its calculable chronos.

The texture of the crisis may bear little resemblance to that which was anticipated by Paul, nor is it likely that his letters will deliver us the answer. However, the intertextuality that has mounted largely outside of theology proper as "political theology," and which circulates now almost irresistibly back through the apostle's epistles, may help us to see through the schematism of our world. It has, even in theology, made at least preliminary contact with the nonhuman life contracted together with the human.[14] By self-questioning, may we answer to the multifarious lives of the earth? Lives endlessly, provocatively, human come entangled in the unfathomable immensity of the nonhuman: the creatures all—*differently* and *without exception*—gathered, contracted together with each other. Each therefore is gathered within the earth, as part of it and so precisely *as* it. Even if, as imagined in some anthropocene technodreams, we can be put *en masse* on a spaceship and separated from the planet, we would remain earthlings.

The earth names the space of this political theology, then, not as a flat surface or a hard ball "on" which we live. The earth will not smooth down to that globe taken so for granted in the spirit of sovereignty and oblivion. It is not yielding submissively to the religio-politico-economic schematisms of what we may call *anthropic exceptionalism*. Earth names not matter beneath us, not a space lying static beneath time, but the teeming sphere of our collectivity. It pulses with the polyrhythmic temporalities of a planet embedded in a cosmos itself multiple and vibrant beyond our imagining. As Nicholas of Cusa, apparently the first within the Christian world to teach the infinity of the cosmos, also recognized, that cosmic All contracts itself to each creature: "the universe is contracted in each actually existing thing."[15] And, we add, with the help of a more current cosmological schema, such a contraction of the cosmic environment to the particular creature takes place in the now-time of each event of becoming, each "actual occasion or "drop of experience.").[16] We will consider later how every microcosmic space-time happening, excepting no creature human or other, takes on social, and therefore in a certain sense political, significance. How would then the eventiveness of every becoming pertain to the kairos-event of crisis and novelty?

And we will wonder, in view of the earth crisis that the *anthropos* embodies, if any kairos can arrest and suspend the chronos of planetary doom—enough, at any rate, to trigger the realization of another possibility. Surely not, we suspect, if we await some messiah, as great exception to our condition, to come do it *for* us. And yet what we—we the collective of human earthlings, in this instance—are doing for ourselves has just about run us out of time.

But not quite. And, after all, deadlines, I've found, can from time to time spur surprising actualizations.

THIS POLITICAL THEOLOGY

To read together our collective earth moment, we will spiral in this insistently condensed exercise through an account of the political, of the earth, and of theology. These three chapters that make up *Political Theology of the Earth* link political philosophy with ecology and theology in order not just to theorize, but to agonize and to mobilize. Without anxiety.

The political signifies the social schematism of a contraction: it is the gathering and being gathered together of humans, beyond tribal cohesion, in the *polis, civis*, the urban units of *civi*lization. And, as Wayne Meeks' *First Urban Christians* classically demonstrated, the movement that came to be called Christianity has, since its rapid urbanization by Paul, always been *political*. Its ancient antecedents, largely imperial, along with its Jewish origins, often counterimperial, were always already political in their theisms. Gradually, Christianity would disperse its shifting theopolitics across its known cosmos, indeed its cosmopolis.[17] Its materializations have been as schematically diverse and internally conflicted as its civilization.

In our time, the world, the schema of our coexistence, is rapidly being degraded by its urban elites into a planet of slums and dumps.[18] Inside or out of whichever ambiguously vibrant metropolis, no heirs of either testament—indeed of any spiritual tradition, whether cosmopolitan or indigenous—can responsibly ignore the material schematism of economics and ecologies now so densely contracted in the polis. The biblical chain of urban signifiers hangs heavy, just beneath attention, on the Western political imaginary: Babel, New Jerusalem, Whore of Babylon, City of God, City of Man, City on the Hill . . .

If religion has never been apolitical, still, even as *political*, it is never simply identical with *politics*—that is, with the structures of the state, and so with institutions that religious practice may shape, sanctify, question, or protest. *Theology* names a religion's theoretical practice (that *theo* does double work). So then theology is not politics, but it is *always already political*.

At the same time, but by an inverse theorization, *politics is always already theological*. And here we cannot avoid the practice of what across many disciplines is called political theology. Nor do we evade its pivotal, problematic text, *Political Theology* (1921), the short volume by the legal theorist Carl Schmitt, whose central postulate states itself clearly: "All significant concepts of the modern theory of the state are secularized *theological* concepts."[19] This is a shocking hyperbole for any who consider "the secular" to be purely the creation of modernity.

The alternative to ahistorically presuming secularism's creation from nothing might then be to recognize the secular as *the secularized*. Factoring temporal process into secularity itself surely makes sense. After all, the very word *saeculum* signifies a "time," an age or epoch, indeed a schema. Schmitt, analyzing how political concepts "were transferred from theology to the theory of the state," is focusing on the modern concept of sovereignty. Thus resounds the privileged example of the transition of sovereignty into modernity: "The omnipotent God became the omnipotent lawgiver."[20] Schmitt offers the juridical analysis of *Political Theology* precisely not as a work of theology but of sociology. The hidden theory of the *deus omnipotens* in modern politics is thus exposed by a secular social theory.

It is Schmitt's deployment of the phrase *political theology* that has fueled the recent outburst of political theologies across multiple disciplines—most of them avowedly secular, most of them *not* theology. Schmitt did not invent the phrase. He deploys it in mockery of a mockery. It had been the premier anarchist of the nineteenth century, Mikhail Bakunin, who coined it in ridicule of fellow revolutionaries who were partially motivated *politically* by their Christian faith or Jewish practice. Religion, for Bakunin, was itself the original sin. He wrote "The Political Theology of Mazzini and the International" (1871) in order to expose and deride the lingering God of that leading Italian revolutionary.[21] Without much discussion of Bakunin, Schmitt's *Political Theology* nonetheless concludes on "the odd paradox whereby Bakunin, the greatest anarchist of the nineteenth century, had to become in theory the theologian of the antitheological and in practice the dictator of the anti-dictatorship."[22] Schmitt's ironic deployment of Bakunin's inversely ironic phrase may tip the present project into a perversely tripling irony, performing the theological antitheology to the nontheologian Schmitt's own antidote—his dictatorial theology of omnipotence as the boost to sovereign power.

We may hope to elude dictating yet another round of antidictatorship. But I see no way out of the satirical chain reaction. This means that the present text may no more escape the logic of Schmitt's postulate than do the plethora of theorists on the postmodern left, largely atheist, who oppose Schmitt's counterrevolutionary politics yet take with utmost seriousness his political theology. From *A*gamben to Žižek—the chain reaction won't stop—they happen also to be the non-Christian fans of Christ's fan Paul. They engage Schmitt, but do *not* constitute an ideological fan club. Since most of these new proponents of political theology are no more theologians than was Schmitt, the present author (undisguisably a theologian) has let this book take its present name.

Indeed, if political theology did primarily designate a room in the house of theology, it would be at the present moment of little use to theology itself. Under various other names, the inherently political animus of progressive religion has long, at least since the Social Gospel movement among Euro-American and Afro-American theologians of the late nineteenth century, sought zones of resonance and intensifications of solidarity with expressly secular social movements.[23] In its defining disputes with orthodoxy, this interactivity builds upon the interdisciplinarity of the entire modern history of liberal theology. Given the statistical shrinkage of the mainline base for such liberal/progressive religious traditions (yes, our time is contracting), it seems politically futile for these forms of Christianity just to keep proclaiming our own modern or postmodern or postsecular relevance. But might we welcome the recognition by political theorists of a largely hidden theology always at play, for good or ill, *within* the political? Not that such recognition will somehow shore up the cultural prestige of theology and the mighty fortress of its God. But it does open a vein of vivid transdisciplinarity in which theology itself may offer politically useful transcodings between the religious and the secular. Of course, then, that political recognition will be complicated by the difference between theologies and therefore between the forms of their political secularization.

Theology here articulates what unconditionally matters. Within its Abrahamic materializations, it takes up the schematism of "God" but recognizes that *theos* as *one* way of naming the unconditional condition of all that is. It therefore invites the practice of "comparative theology," which embraces nonconversionist interactivity across the whole spectrum, East/West/North/South, of religious/spiritual practices.[24] Without the help of theologically deft forms of religious and cultural pluralism, attentive to the multiplicity of materialization, layers of contradiction, and modes of hiddenness of theology itself, the very notion of political theology remains a Schmittian mockery.

If political theology as such has not developed until quite recently as a branch of theology, it almost did. The phrase had been resolutely reclaimed by theologians of the post-Holocaust German left starting half a century ago. In a movement of European solidarity with the fresh voices of liberation theology in the global South, Johannes Metz, Jürgen Moltmann, and Dorothee Sölle emancipated the phrase from the Nazi collusion with which Schmitt had tainted it.[25] John B. Cobb Jr. built on their work in *Process Theology as Political Theology* (1982). The work of each of these four theologians remains greatly influential, but the phrase did not quite take off in theological circles. It seems that *political theology* had felt too Eurocentrically generic to catch on among progressive Christian thinkers for the rest of the century. At least in the United States, the phrase perhaps threatened to dilute the bursting particularism of liberation, Black, female, gay, lesbian identities. It seemed too prematurely universal in its humanity. And for ecotheology, it signaled too anthropocentric a *polis*. These were not so much disputes as failures of resonance.

Now, in a different time, political theology returns to theologians largely by way of our inextricable involvement in philosophy and in social activism as well as in recent political theory. This does not mean one has recourse to political theology in order at last to transcend the identity politics of the religious (or any other) left. I hope rather to think together with any who work to gather a de-essentialized, dense—indeed contracted—entanglement of our differences. It takes constructive theological form in what Moltmann has in the present century, mindful of growing grounds for pessimism, called "the solidarity of hope."[26] Those philosophers of religion who offer now a "radical political theology," poised between the death of God and an indeterminate future, may invite a less sanguine solidarity, but they set forth the "insurrectionist manifesto" of a radicalized democratic possibility.[27] The alternative then will speak or unspeak its theos as needed. Inasmuch as it provokes a vigorous alternative to the Schmittian concept of sovereignty, it will insist, as Jeffrey Robbins writes in the name of a radical democracy, "upon the immanence of our common life together and the generative power that comes from our modes of cooperation, both already present and still to come." For "this is a project that is theopolitical as well."[28] And, within ecotheology, attention not just to the capitalist drivers of climate change but to their political schematism has led Michael Northcott, for one, to an intensive engagement of Schmitt and thus—in an extraordinary collation of political theology with ecotheology—to an answering "revolutionary messianism."[29]

The point is not, I repetitiously underscore, that theology is only recently waxing political. There is little historic theological thought that does not locate itself in some aftercurrent not just of Paul's urbanity but of Augustine's "two cities."[30] Nor has it just now turned progressive, as the late nineteenth-century Social Gospel movement demonstrates, nor only recently learned to theorize the multiplicity of urgent social issues. Ecofeminist and ecowomanist theologies of relation, along with the philosophical theology of process, have for nearly half a century constructed transdisciplinary intersections between the political, the ecological, and the discourse of the ultimate.[31] They have done so with respect for the secular boundaries of religious discourse—in part because they share much with secular critique of religious, as in mainly Christian, overreach.

Just as a convenient example, I persisted in theology partly because of early exposure to the work of Paul Tillich, his war-intensified kairos, his God who does not "exist but is the ground of existence," "Being Itself," and his inspiration of Mary Daly's verb version, "Be-ing." At almost that same mid-seventies moment, I came into the force field of a Christian deployment of Whitehead's philosophy of process, its take-down of omnipotence, its radical relationalism fomenting John Cobb's early warnings about the global ecology and its "religion of economism" that has the world in its grip.[32] Cobb taught that the founders of the great ways—Moses, Socrates, Confucius, Laotzu, Jesus, Mohammed—were secularizers *avant la lettre*, turning their communities beyond "religion" and toward their social world. He distinguished this "secularity," tuned to the challenge of social justice in the *saeculum*, the age and its politics, from "secularism," which is itself just another religion.[33] These strong influences combined and recombined, in me and in many, with the exploding *kairoi* of sex, gender, race, class, and species politics. But the wider influence of process connectivity was limited. If sometimes still playing a zero-sum game with each other, the plural identities of liberation found deconstructive destabilization and so fresh interdisciplinary refraction through poststructuralism in the later decades of the last century. And the pluralism of Whiteheadian thought, now entangled in a Deleuzian rhizome and a fierce ecosocial planetarity, does not cease to seek out fresh publics secular, religious, and spiritually indeterminate.

These prophetic traditions of progressive theology are not one. And this brief account offers only a personal contraction of the dramatic shift in the texts of theology from the late 1960s on. But one might generalize to say that these theologically yoked ecosocial justice experiments have without exception offered their thought back to their respective religious legacies. One may say also that they have done so in the spirit of an autocritique that means to mobilize activist transformation within those traditions. So, for instance and concretely, the struggle to overcome entrenched habits of masculinist and heterosexual normativity continues amidst multiple religious institutions. As do waves of insistence, against liberal complacency, upon renewed work against racism within and beyond our own institutions. Always, however, the prophetic movements have aimed beyond religion—as, for example, the liberation theologians and their transformative, base Christian communities of Latin America accepted extreme risk as the cost of challenging the U.S.-backed dictatorships in the 1980s.

In its explicit embrace of the social justice ethos and its creation ecology, theology is only recently picking up the dropped discourse of political theology, with its Schmittian baggage. I see no problem with this delay. This emergent discourse will not deliver redemption, only some timely transdisciplinary insight into delays that destroy. The postponement that has become acute since more or less the first Earth Day of 1980 is now subject to the new delay—the one that requires a language of "aspirational fascism,"[34] or "authoritarian populism," with its avenging racism deployable against both citizens and migrants, its climate denialism directed against the earth itself. This double politico-ecological deferral of justice speeds us toward planetary catastrophe: we are at once stalled out on hope and rushed toward the end. The double-talk of "fake news" greases the rails of progress toward doom.

In this double-time the deep intersections of injustice and unsustainability expose themselves, as this book argues, in an extravagant exceptionalism. Its absolute investment in the extractivism and exterminism that pump the global economy works in tandem with the American exceptionalism that lets us (U.S.) use or abuse the global at "our" sovereign will. If that will has recently embodied itself in a persona volatile to the point of derangement, is it more than a pumped-up rearrangement of a long-term socioeconomic schema? It effects new unification through really old racial, sexual, and class resentment. Yet—and this may prove fortunate or catastrophic—there appears to be as much instability as there is strength to the schemes, conflictually national and global, joining the political to the economic.

Any certitude of analysis seems now certainly dishonest. Certainty itself exercises a sovereign simplification by which it unifies its knowledge and so its world. Its very modern world: after all, "knowledge is power"—by which Bacon meant a certain kind of knowledge, i.e., certainty. Certainty allows the conquest of the other, which it simplifies as object. At the same time, the sheer multiplicity of interconnected threats belies any monocausal simplification—whether political, sexual, racial, economic, ecological, or religious. The unconditional itself does not deliver certainty. Uncertainty, however, does diminish responsibility; it multiplies possibilities for response.

DARKENING HOPE

There is an indiscernibility between the darkness of what we cannot know and the darkness of what we do not want to know. A shadow formed of two darknesses: that which exceeds our capacities to think and that which feels ethically unthinkable. Theology, as it turns out, has an ancient practice for liberating insight from certitude, for thinking at the edges of the unthinkable. Called negative theology, or *apophasis*—"unsaying"—it is born in antiquity as a negation of any name, dogma, or knowledge of the divine, however true and nonnegotiable it may seem. The unconditional twists always into the unsayable. Apophatic theology operates as a practice at once theoretical and spiritual, a means of mystical insight.

Such negation of any finite certainty as to the Infinite now takes on the shades of another negativity, that of an ethical critique, which shades readily into pessimism (as, for instance, the brilliantly tuned darkness of "Afro-pessimism"). The mystically unthinkable links up to the ethically unthinkable; the apophatic abyss morphs into the ecopolitical horror. So the darkness shadows any hope for a viable future. But rendered theologically, that "cloud of the impossible" (after which I named a book, in tribute to a much older cloud) embraces its own multiplying negations. Uncertainty undoes all optimism. But it may, for that very reason and in our circumstances, enable a "hope in the dark."[35]

So we might insist that, as theology itself reenters the fray of political theology, it do so in mindfulness of its cloud. I had threaded theology to a political ecology of "apophatic entanglement." Such theology keeps exposing the creaturely vulnerabilities of our inescapable, indeed our nonexceptional, interdependence, relations themselves multiplying and deepening unpredictably into the unknowable. At one cosmic angle the unknown darkens into metaphors of divinity. Yet the unknowable at an earth tilt shades into the unbearable, even as populations of human and other creatures lie exposed to unthinkable but quite probable depredations and demise at the planetary scale.

A political theology of the earth casts always and mindfully this shadow of many darknesses. Therefore it can sometimes appear as a "negative political theology." It will let us practice a systemic mistrust of certainties, however well intending. Only so can it tender courage for the thinking of the unthinkable, epistemic and ethical. Recognizing the opaque lining of any certainty , we think not less but better. We find clues in dark places.

So in this book we will reconsider Schmitt's concept of a sovereignty decided "in the exception." We may find thereby a clue to a whole current schematism of exceptionalisms. And it is precisely Schmitt's implication in the fascism of his time that lends a certain insider insight, a glimpse into the shadowy inside of the different but not unrelated proclivity of this, of our, time. When is that? As I write it is time passing through its proto-fascist farce of the bully sovereign. I hope by the time you are reading this—the time of books has a slowness—the possibility of fascism has receded. In any case, the precarity of our democracy will not have. In this political peril, the U.S. is in fact hardly exceptional, but joined by numerous other nations suffering differing but related temptations to a white-supremacist authoritarianism. And the high probability of coming great waves of climate migrants, millions upon millions fleeing the waves of flooding coastlines, or the heatwaves of spreading drought, added to and intensifying migrations forced by war, threatens to bring out the worst in host democracies.[36] It will also, here and there, provoke the best, and in reaction, *worse* worst, and perhaps, better again, who knows . . .

But where that chain ends is in apophatic ellipsis. In other words, responsible ecosocial predictions may brace themselves with sound science and unfake facts. But they do not follow a teleology of hope or of doom. And, as *theology*, this political theology does not perform the Messiah or announce her coming. Indeed "the messianic thwarts the teleological unfolding of time (the Messiah will never appear in time)."[37] That is Judith Butler thinking with Walter Benjamin. The pretensions of a predictable chronos fail us. They do not come in time. But neither is this Jewish messiah timeless, enthroned beyond time itself. The messianic, perhaps then even Paul's messiah, called *ho Christos*, is always "coming"—never captured in a final revelation of gore and glory.

If the moment of crisis is to open at its edge, its *eschatos*, as kairos, it seems that a specter of "messianicity" (Derrida) starts to materialize within it. A historically worn yet inexhaustible possibility shoots—like the shuttle of the loom—through the tangles of impossibility. It might make up in creativity what it lacks in power. Who knows?

PRECAPITULATION

This political theology of the earth will weave its way through a threefold argument by way of a schematism of three chapters: on the political, on the earth, on the theology that here gathers them. Chapter 1, "The Political: Sovereign Exception or Collective Inception" takes on the locus classicus of political theology as such, Schmitt's politics of friend vs. foe. Its unifying antagonism is answered by both William Connolly's and Chantal Mouffe's notions of "agonism." Such antagonism is hard to miss in current politics of we vs. they. If, for Schmitt, sovereignty is established "in the exception," the emergency that displays the exceptional power of the leader, it lives from the secularization of a theology of omnipotence. We consider Kelly Brown Douglas' genealogy of white exceptionalism, correlating it to multiple registers of exceptionalism trending toward planetary emergency. The now-time leads through its dense Jewish/Pauline intertextuality toward an alternative theology for the political present.

The planetary time of crisis registers as climate change. Any responsible politics now faces—or denies in plain sight—the effects of a theologically sanctified anthropic exceptionalism. So chapter 2, "The Earth: Climate of Closure, Matter of Disclosure," pursues its time-sensitive matter (the data always worse by the time of publication) with counterapocalyptic intent. A poem of the great African American ecopoet Ed Roberson—"To See the Earth Before the End of the World"—will guide us. Reflecting on the meltings and the floodings, the droughts and the fires, the immigrations and the inequities already materializing in the Anthropocene, Donna Haraway and Karen Barad invite us on some energizing loops through the nonhuman.

If the chance of an ecosocial inception depends upon a shift of political theology, we must then actually reconsider the matter of theos. For, after all, theology itself seems to be failing right along with democracy and ecology. Chapter 3, "Theology: 'Unknow Better Now,'" meditates on a theological unknowing, apophatic theology, in relation to the standard certitudes of a theology of Christian exceptionalism. Process theology poses here the metaphors of a constructive alternative in which sovereign omnipotence gives way to a depth of creative indeterminacy. There an amorously struggling alternative, a *seculareligious* political potentiality, begins to materialize.

As we get free of anxiety, the kairos fills fuller. Its becoming now, in a textured density of intersecting relations, does not determine a future. It overflows the present as possibility. Possibility, that is, for different (possible) futures. If that kairos stirs hope, it works against the chronic hopes of supernatural salvation, secular progress, or any other chronology of optimism.

What does it work *for*? For the now-time cannot open as the mere reaction of an against. How shall we name its dark chance, barely pronounceable as hope? The potential for . . . how do we find words that don't say too little—justice, sustainability, an ecosocial public, social democracy? Or too much: a coalition intersectionally dense and vast enough to interrupt the death spiral of climate change and the politics of a capitalism that, in its globalism of growth, is fomenting endless raging, racist, heteromasculinist disappointment. Such coalescence energizes not just resistance but an insistence of that assemblage upon political alternatives? Falling short of fulfillment, such mattering experiments, however local in space and fragile in time, foster a just and sustainable enough common life. Common to the point of the "undercommons," darkly bound together in the promiscuous solidarities of difference—in the precarity of our conditions, this work matters unconditionally.[38]

Too much and still not nearly enough. Too little and still: a beginning.

1

POLITICAL

Sovereign Exception or Collective Inception

Deep in the summer of 2017, an article was published by two close associates of Pope Francis warning of the "political Manicheanism," indeed of the "apocalyptic geopolitics," of recent U.S. history.[1] Observing a new merger of politics and Christian fundamentalism (both evangelical and conservatively Catholic), it traces the rhetorical axis from the post-9/11 mission to "free the world from evil" to the forty-fifth U.S. president's "fight against a wider, generic collective entity of the 'bad' or even the 'very bad.'" The authors of the piece might have added that, beyond the abortion cue, minimal signals of Christian meaning and morality were now needed to gain the white evangelical sanction; the unifying power of antagonism—against Blacks, migrants, and Muslims, against environmentalists, against "her"—sufficed unto triumph. Against the bad *them*, an *us* edged with swastikas and white hoods began again to show itself. Apocalyptic geopolitics surged in the nuclear play with North Korea and, in a gift to the religious right, declared Jerusalem Israel's capital.

It is tempting to move directly with Pope Francis to his operative political theology, counterapocalyptic and *eco*geopolitical. For him, Jesus' down and dirty foot-washing practice symbolizes the alternative to a politics of fear and enmity, of colonization and capitalization. An earth-minding political theology has had no more globally influential a friend than this advocate of "care for the common home."[2]

We might then congruently define the political as *gathering for the common good*—and so signify a planetary commons,[3] a commonwealth, a collective good wrought in the struggle for the common, not in enmity against "the bad." The Pauline kairos of *sunestalemnos*, the contraction that collects, that does not merely shrink time, begins to reverberate.[4] So far so good.

And yet, by itself, such a definition might lose track of a critical dynamism. The political remains a different sort of gathering together, of contraction, than that, for instance, of a family, a community, a religion. These are all shot through with politics, may inhabit the polis, but they do not constitute in themselves "the political." It cannot be thought apart from *struggles with acute difference*. Such difference may pose itself as a danger to the common good—a challenge, a change, or indeed an enemy. The response may be Franciscan and foot washing rather than Manichean or apocalyptic. It may prioritize electoral politics or social movements. But the response only counts as *political* when it gathers its public—as a "we"—in and through shared struggle. With time contracting, what sense of collective struggle organizes a political theology of the earth?

FRIENDS 'N' FOES

And who are *we* anyway? The question of the we—when it becomes a question—appropriately politicizes every engagement (from Tanto's postmodern "What do you mean 'we,' white man?" to the reading "we" to which "I" regularly appeal). Who is the political subject? How does struggle collect it?

Here we rejoin the founding conversation of political theology. Carl Schmitt had in the 1920s defined the political in terms of the formation of a unitary public subject. That unification of people as a people takes place only to the extent that they share a common enemy. In his 1927 classic, *The Concept of the Political*, Schmitt posits that "the specific political distinction to which political actions and motives can be reduced is that between friend and enemy."[5] Social identity is formed as a unity achieved through conflict. The political "we" in this reduction is produced by a shared antagonism against some Other, some "they." This is the simplified collectivity of unification, not mere coherence across difference. Any more peaceable and pluralist notion of the political falls to his critique of the weakness, bureaucratism, and failure of democratic liberalism.

The political philosopher Chantal Mouffe argues, however, that Schmitt here poses a "false dilemma": Either the heterogeneity leads to "the kind of pluralism which negates political unity and the very existence of the people," or, she says, "there is unity of the people, and this requires us expelling every division and antagonism outside the people."[6] Schmitt insists upon the latter. The exclusion signifies less a once-for-all purge than an ongoing work of unification. Yet "the people" cannot be sustained through free discussion but only through the decisive exclusion of what counts as divisive.

Brushing off the Christian counterposition, Schmitt insists that the biblical injunction to "love the enemy" has nothing to do with political enemies: "It certainly does not mean that one should love and support the enemies of one's own people."[7] That exegesis immediately opens the politically theological door to Schmitt's definition: "The political is the most intense and extreme antagonism, and every concrete antagonism becomes that much more political the closer it approaches the most extreme point, that of the friend-enemy grouping."[8] This affective intensity in its "extremity"—unlike liberal rationality—here fires up and fuses together its unified political subject. We all recognize the power of a common enemy to galvanize the "we." Many of us learn the affective force of this negotiation from the prepolitics of the family, where a fragile harmony would be instantly restored by assent to a common enmity, often racialized.[9]

"Tell me who your enemy is and I'll tell you who you are."[10] Hard to dispute. But then any common good lives thus as parasite upon the common enemy. Of course for any (nonfundamentalist) politically engaged theology, it is tempting to read the Schmittian politics of friend versus foe as itself the foe. *Our* "we"—as in, for example, *we* friendly Christian progressives—wants amity, not enmity. (Signs proclaiming, "Love trumps hate!" in childlike scrawl sprouted up overnight at seminaries like mine on November 9, 2016.) But here's the catch: I cannot see how to deny that one who so nakedly embodies this hatred is properly called an enemy—indeed "our" enemy, dear scrawlers. More important, neither can I deny the chance that such shared enmity can gather a critical mass of us into an affectively intensified solidarity. This enmity vs. enmity involves a coalitional togetherness that is not simply *for* love, justice, and a more common good. Its solidarity arises, at the same time, as militantly *against* the racist/sexist/denialist regime.

"Love your enemies"? Sure, amen. You don't love to hate them, or else you mimic them. The difficult love called *agape* presses hard against enmity itself. It might soften and shift it. It nonetheless recognizes the enemy as such. And sadly that agape does not guarantee that your enemies then will stop performing as enemies. They might just crucify you. The work of love for a more common good devolves readily to a politics of sentimentality and evasion. .And yet as a secular "politics of love" (Michael Hardt and Antonio Negri), as "distinction amidst relation" (Marcia Pally), or as an interfaith "revolutionary love"—it can tap in itself a political, indeed a geopolitical, love struggle.[11] And fighting *against* current symptoms of systemic enmity, we are nonetheless fighting first and last not against but *for and with a public*. Otherwise we struggle only reactively, not proactively.

Biblically speaking, the Second Testament's intensified love mandate is an iteration and a contraction of the first's, which also exceeds the love of "neighbor" and requires love of the immigrant, the foreigner, the stranger, those others so readily cast as the enemy: "You shall also love the alien, for you yourselves were aliens in Egypt" (Deut. 10:19). In Schmitt's version of Christian political theology, however, neither alien nor foe are to be loved; rather, they are smoothly collapsed into one antagonistic Other. He defines the "enemy" as: "the other, the stranger; and it is sufficient for his nature that he is, in a specifically intense way, existentially something different and alien."[12] The enemy need not be doing anything wrong, mounting an attack or threatening a war. So one grasps how in the 1930s this widely respected thinker could support the construction of the Jew as unifying enemy. If the political collective must come down to an oppositional *identity*, the religio-racial identification of an ethnic other becomes the most effective means to essentialize at once an enemy and, therefore, as its immediate effect, a we. A we-One versus the they-Other.

The designated aliens of religion and unwhite race may shift. The strategy, however, returns in force. But how then would *we* (those of us who share my enmity against the purveyors of enmity) avoid playing a mirror game of antagonisms? How are we not forming our political identity in opposition to the identitarian oppositionalism of our foe-producing foes? Indeed, don't we, for the sake of solidarity and political impact, form identities also in opposition, first to those who oppose us but soon to all who do not share our precise angle of rage (sex, gender, race, class, species . . .)?

If, alternatively, we avoid hostility, we readily fall back upon a disempowering civility, an inclusivism invested in the liberal rationality that Schmitt, not without reason, deemed a failure in the 1920s. (He had the Weimar Republic under the crushing treaty of Versailles in view.) A certain civility surely forms the very basis of the civic—of the polis, of the political as the context of public struggle. But what if that civility has undergone a fusion with the rationalizing liberalism of presumed progress, and then with the neoliberal rationale of global capitalism? No wonder it has become increasingly ineffectual in forging its own democratic *we*. *Our* internal contradictions—between radicals and moderates, between cultural identities and economic classes, between secularism and religion—fester beneath the surface of a limply presumed consensus.

OR AGONISM

Chantal Mouffe articulates the third way, beyond liberal consensualism and mere antagonism, of a "democratic agonism." *Agonism* signifies not enmity but struggle. Taking seriously Schmitt's critique of the liberal political expectation of rational consensus, she argues that "a healthy democratic process calls for a vibrant clash of political positions and an open conflict of interests. If such [a clash] is missing, it can too easily be replaced by a confrontation between non-negotiable moral values and essentialist identities."[13] For instance, among activists drawn honorably and *en masse* to the chance of a more social democracy, one also frequently encounters a confusion between political organization and identitarian safe spaces, where they may be spared the clash of even contiguous perspectives. Then we suffer a dangerous reduction both of pluralism and of the wider solidarities that require it.

The political philosopher William Connolly lifts up as key to a working democracy the practice of "agonistic respect." The concept runs in some ways parallel to Mouffe's democratic agonism.[14] For Connolly, such agonism resists the resentment—religious, social, or economic—that fuels antagonism and its spirit of retaliation.[15] He foregrounds the operations of religion amidst secularity without naming it political theology. Well before the vengeful buffoonery had ascended to U.S. sovereignty, Connolly was warning of the dangers in late modern capitalism of hubris and its corollary, disappointment. He demonstrates in particular how the ensuing affect of resentment fueled the improbable merger of the religious right and the economic elite over several decades in the late twentieth century—the fusion that had so successfully launched its apocalyptic geopolitics in the Reagan epoch.

"Today 'religion'—a flat word collapsing creed, liturgy, and spirituality into one compass—surges into the pores and pulsations of politics."[16] (Many religious thinkers would concur as to the flatness of "religion.") Resentment religiously fueled—against the world, against those who seem to embody it, whether to benefit unfairly from it or to call for responsibility to its future, against those who believe otherwise—energizes the base of a pyramid of capitalist power. Connolly is at pains to demonstrate the potent tension of this alliance: the party elite "discounts its responsibilities to the future of the earth to vindicate extreme economic entitlement now; while the [evangelical base] does so to prepare for the day of judgment against nonbelievers. These electrical charges resonate back and forth, generating a political machine much more potent than the aggregation of its parts."[17] Crucially for the present project, Connolly does not confuse this form of religion with more responsible forms, whatever their beliefs.[18]

The "we" of what Connolly dubbed the "capitalist-evangelical resonance machine" may be now taking a new turn. Its foes may be now designated more by Fox and "friends" than by evangelical, or any "religion's," teaching.[19] But it shows no signs of diminishing political force. A resentment against some readily marked set of fellow citizens who seem to be getting away with something—some economic or erotic entitlement, some menacing criminality or dark attitude—has for decades now been expertly stoked, promising restitution for "us" and punishment for "them." This retaliatory affect fuses with the deeper condition that Nietzsche had diagnosed as *ressentiment* against the fragilities of earthly life, of mortality, and so of time itself, of time's "it was."[20]

Under the best of circumstances—and before climate change—life remains precarious, agonies multiply, and time runs out. Every faith ministers to the uncertainty of future, in ways honest to the unknown. Or not. "Capitalism depends upon faith in the future."[21] Even under the guarantee of a capitalist-fundamentalist schematism, however, the chronology of success routinely fails to deliver. Chronos disappoints. The promised rewards get deferred yet again—and afterlife is never quite enough of an advance. Antagonism mounts. And it will be stoked by convenient narrations of the strange and different as the foe, those jewishmuslimexicanblacklgbtqaliens. Blame them, not your Fox friends, for the precarities of wealth, health, and happiness. Of time itself.

If we took a hint from the Pauline signifier of the kairos, it was because its now-time wants to free us not from time's limits but from the fear of time limits. It contracts in itself a great tension of temporal becomings. Time within times, world within worlds, body within bodies, all contracted within the earth of our own time. Might we begin, for the sake of a political theology mindfully embedded in the fragilities of the earth, to solicit a correlation between an indeterminate kairos and a respectful agonism? Can it politically resist the ever looming correlation between chronos and antagonism?

"WE OWE EACH OTHER EVERYTHING"

Agonism means struggle. So the democratizing ethic of respectful agonism does not anticipate the cessation of struggle. Nor does it depend upon consensus for action. We may infer therefore that it does not erase the torments of time, let alone those routinely inflicted by systemic antagonism. We might amplify—that is, render audible—the relation of agonism to *agony*. But does this honest agonism not open the backdoor into political immobility and the pathway to despair?

Or very differently, it may teach a "public practice of lament."[22] In a remarkable work tracking sites of improbable aliveness in the wake of the horrifying violence that has wracked East Africa, Emmanuel Katongole finds over and over this phenomenon of lament in prayer, poetry, and song: a "complex performance, a discipline, that involves many actions": "crying out, exercising memory, maintaining silence."[23] In his own political theology, he has demonstrated that lament, in the contemporary culture of Africa as in the First Testament's psalms of lamentation, "is a way of hoping in the shattered ruins of life."[24] The sharing of agony does not cover over but exposes a traumatic history. Only so might its participants live again. Affects of anger and compassion effect a solidarity that sustains. And the systemic justice required does not then confuse itself with retaliation.

The struggle to survive a trauma can, in other words, teach us much about the struggle for political transformation. For the refusal to struggle and sometimes agonize together, and often over our own differences, means the repression of the political space in which "a vibrant clash of political positions" can take place. Then the possibility of a shared good, a more common weal, gets defeated in advance. This is no mere problem of the right. The left has failed routinely to struggle faithfully, respectfully, with difference—with internal, neighborly difference, let alone with ideological enmity. Our slogan of "diversity" routinely reduces to a single difference. Then, mirroring the antagonism of reactionary unification by the alien, we defeat our own principle of difference. It is not that democratic agonism offers a straightforward alternative to the machine of antagonism. Distinguishing them under conditions of righteous anger and legitimate resentment poses its own struggle.

One might then define the political, not reducible to politics but not excused from its institutions, as the struggle for a *more* common good. The struggle has always been acute, for the common has always been shoved beneath the political as its dull background of needs, services, and resources. When it comes forward, the common shares the radical sociality, the demanding egalitarianism, of any strong sense of the political collective. As the foot-washing pope—arguing for neither socialism nor democracy but the "integral ecology" of a global common good— takes his cue for the "common" from the poor and the earth,[25] we too take ours from the dehumanized and the nonhuman: the humans forever slipping beneath the status of the human, the nonhumans by definition crammed there beneath all humans. It is emphatically not the common but the exceptional, not the commons but the enclosure, that organizes the globally dominant political economy. Global capitalism stages the precise opposite of a global commons.

Nonetheless we recognize that the language of "the common" can, in the needful quest for coalitional common ground, quickly lose the down-under radicality of its collective, the antifoundational earth of its common ground. The common, in other words, can lose its own uncommon differences. But it turns out that there are entire undergrounds forming in the ground beneath, down there in the dirt with the dehumanized and the nonhuman. Why not then say that a political theology of the earth holds us to a *more* common good, if and only if we hold that expansive commonality to be inextricable from what Black critical theory has called the "undercommons"?

In the interest of "fugitive planning and Black study," Stefano Harney and Fred Moten propose, with uncommon discursive beauty, the undercommons of a "we" who "are disruption and who consent to disruption. We preserve upheaval."[26] Sent "to renew by unsettling, to open the enclosure whose immeasurable venality is inversely proportionate to its actual area, we got politics surrounded."

In the disclosive opening of its enclosure, this undercommons—to the boundless ressentiment of the enclosers—does constitute a virtual planetary majority. Its enclosure enfolds, encompasses, surrounds. Can it be mobilized against its own exclusion? "Politics proposes to make us better, but we were good already in the mutual debt that can never be made good."[27]

The undercommons names a public that has lost hope in politics but that persists in resistance and self-organization. It can hardly be deemed apolitical, even if it refuses any nominal politics. For "it's not the thing that you do; it's the thing that happens while you're doing it that becomes important, and the work itself is some combination of the two modes of being."[28] Does the "thing that happens" evoke the eventiveness of the inceptional kairos? Moten's poetics carries the cadence of a grace without payback. But it not only exposes the pallor of the present experiment, it challenges our proposed agonism: "We are the general antagonism to politics looming outside every attempt to politicize . . . every sovereign decision and its degraded miniature, every emergent state and home sweet home."[29]

That "sovereign decision" can only be referencing Schmitt's concept of the power that is determinate of a strong decider, state, and *Heimat* (home sweet race). Yet the decisive force of sovereignty derives precisely from antagonism—and so, as will become clear, from the "decision in the exception," which is to say, in the confrontation that the sovereign deems to be an emergency. This looming counterantagonism that resists the sovereign antagonism and its democratic masks may gather a needed force. For the present political theology to remain indissociable from the undercommons does not mean, however, to appropriate its antagonism or to claim its space so much as to participate in its struggle. Otherwise the difference between respectful agonism and disruptive antagonism could itself become a point of dissociation, by which the sovereign divides and conquers, degrades and indebts. How wasteful.

We might instead recognize that a political commons (not a politics) worth gathering will respect—as a criterion of our agonism—the "general antagonism." Let no white voice of "we" dilute it. Otherwise the contraction of the Corinthian *sunestalemnos*, the ingathering (the 1 Cor. 7 term) shrinks to a mere retreat from the moment, a pale miniature, not a darkening intensification. Yet this very recognition can perhaps only take place in the spirit of what womanist ethicist Emilie Townes calls "methodically destabilizing antagonistic binaries."[30]

The political agonism with which we are concerned knows itself as a matter of timing. Against the time of a calculable chronos directed by the sovereign enmity and its enclosures, it takes part in a destabilizing kairos struggling for its indeterminate dis/closure. Perhaps the indeterminate mobilizes before and beyond any sovereignty, however democratic; which might mean to exercise the political without capture by politics. Hear a bit more of the prophetic *Undercommons*: "We owe it to each other to falsify the institution, to make politics incorrect, to give the lie to our own determination. We owe each other the indeterminate. We owe each other everything."[31]

If "we owe each other everything," there can be no payback time. There can be the recognition of endless and demanding indebtedness. Even in the face of the monstrous inequity of lives liquefied as commodities, of an economy originating in exploitation and extraction, of the enslavement of the dehumanized along with the nonhuman, of endlessly unpaid white debt. "When we talk about debt, to talk about the unpayability of debt is not to fail to acknowledge the debt."[32] The indeterminate, in a temporality that cannot be controlled and corraled by any institution, disorders every determinism, biological, historical, apocalyptic—and certainly political. Perhaps the political as incorrected, as incorrigible, resists the sovereign "we" (royal, totalitarian, or democratic in whiteness) of politics. Thus Joseph Winters, arguing for a sense of political responsibility shot through with tension and received in melancholia, offers *Hope Draped in Black*. Like Katongole's birth of hope in lament, like the darkness of the not yet knowable, the black drapery demonstrates not enclosure but layered, mourning, improbable possibility.[33]

Might a mindfully indeterminate and interindebted collective of color, of endless colors, thereby and against the odds organize in itself the earthen (under)commons of a planetary public? It "has politics surrounded." It emerges, if it does, not as a phantasm of universal freedom but in the contingency of this mutually owed everything. Nothing to bank on. Which might be something to act on.

CAPITALIST THEODICY, CRITICAL DIFFERENCE

The determinate is rapidly reneging on its promises. Our chance, if we have one, lies in the indeterminate. Yet in the chaos of its multiplicitous *anarchos*, its resistance to the one *arche* that is the origin of every sovereign order, there is no lack of organization. Self-organization among humans and nonhumans alike is the activity of assemblage, of the production of ensembles of interactive difference—like a cell, an organism, a jazz band, a revolution. Its self-organizing complexity subverts not order as such but the unifying simplification.

We can now therefore nuance the earlier characterization of the political as the struggle for the common good. We might define the political this way: as *collective assemblage across critical difference*. Critical difference signifies the crisis that difference effects for an emergent public, the divergence that demands fresh acts of self-organization. Diversity is then not a democratic diversion but collective diversification. The self-organization of a public will falsify any institution that obstructs it, that does not find itself in the indeterminacy of its own history.

Self-organization in the sciences signifies the open process in which complex systems emerge, often from emergency, always at the "edge of chaos." The edge of chaos refers here to a phase transition in emergent systems; complexity requires order but not too much of it, in evolutionary as well as revolutionary process.[34] For a theology offering itself up to better secularizations, this is one reason to read the genesis process, always already political, as something alien to the story of an all-determining sovereign creation. As the next chapter will elaborate, there appears, instead of that top-down creation, an ancient narrative of self-organizing complexity at the edge of chaos, indeed on the "face of the deep."[35] It reads with willful anachronism the racial displacements of the "darkness on the face" of that *tehom,* that feared depth—the oceanic undercommons of the theology of creation.

Connolly solicits the science of complexity for political theory in the face of the crucial difference that global warming is currently making. For a universe "far from equilibrium, nonlinear and full of irreversible processes," fragility is not an exceptional emergency.[36] Connolly opens *The Fragility of Things* with a meditation on the earthquake of Lisbon of 1755 and on Voltaire's satire on the theological justifications of unthinkable tragedy—that is, on the faith in omnipotent determinism. Theodicy, or the "justification of God" in the face of evil, serves Connolly as a parable of the self-justifications of neoliberal capitalism—of its political theology, one might say. Whatever sacrifices are imposed upon the poor and upon the earth are for the "best of all possible worlds" (Leibniz). Such a political theodicy "acknowledges many evils and treats them as necessary effects of impersonal markets when the markets are allowed to rip and irrational state interference is not allowed."[37]

No longer pretending even an ultimate trickle down to some common good, this stage of capitalism nonetheless continues to justify its own nondistributive enclosures with a modern time line of progress. To be taken on faith, secular or fundamentalist.

Neoliberalism treats the multiple self-organizing ecologies of the earth—geological, biological, climatic—as externalities irrelevant to its own organization of the world. This world scheme with its extractions, exploitations, and extinctions ignores the fragilities of the world in which first the most vulnerable but soon the all of us will be in peril. Precarity, rather than the good, is getting common.

ETHOS OF INTERCONNECTEDNESS

Warningly, Connolly had evoked in 2013 Karl Polanyi's 1944 analysis of how the "Great Depression unleashed or intensified fascist movements in the Euro-American world."[38] As Polanyi had written in *The Great Transformation*, "the fascist solution of the impasse reached by liberal capitalism can be described as a reform of market economy achieved at the price of the extirpation of all democratic institutions, both in the industrial and in the political realm."[39] Then, months before the 2016 election, Connolly amplified Polanyi's prescience with his own. Regarding the potential for an aspiring new fascism, Connolly comments that it would update the older evangelical-capitalist resonance machine: "The old machine will not be replaced, then, but transfigured, with a few old free market priorities jostled to mix an intensification of nationalist, supremacist, protectionist, and Christian forces more explicitly into it."[40] This transfiguration illustrates his earlier remarks upon the metamorphosis of the merger of right-wing politics and religion. Connolly's attention to fragility becomes all the more poignant in light of the U.S. electoral breakthrough of what he since has called "aspirational fascism" in its manipulation of the race, class, and masculine insecurities heightened by the very logic of neoliberalism.[41] While its economic system postulates its mode of self-organization as an "impersonal market rationality," independent of other systems, and attacks "big government," it depends upon the political force of the state to promote, protect, and expand market processes, maintain corporate personhood "with unlimited rights to lobby and campaign," to police labor unions, to prevent action on climate change, etc.[42]

Connolly insistently reveals the economic concealment of the multiplicity of self-organizing systems in which our political economy, indeed our life, is embedded: "a cosmos composed of innumerable, interacting open systems with differential capacities of self-organization set on different scales of time, agency, creativity, viscosity and speed."[43] He is close to new materialists, like Jane Bennett, but rare among political philosophers in his attention to the interdepending vitality and volatility of the multiple self-organizing systems, economic but also ecological, in which our political self-collecting takes place. The systems remain indeterminate, but to different degrees. Time does not express a pure openness, freed of historical effects. Rather the past repeats itself in the present, but ever differently and never just predictably. Here Connolly builds on A. N. Whitehead, whose philosophy of process infuses the present theology.[44] And though Connolly does not fly the flag of "political theology," religions and their theologies persist in his lists of these complex systems, exercising influence conservative or pluralist, reactionary or democratic, direct or hidden.

Indeed, because a "fungible element of mystery" persists amidst the indeterminacies, his philosophy "welcomes exchanges with theologies seeking to engage the latter's adventures even as it respectfully contests some elements in this or that version."[45] These complex human and nonhuman systems intersect sometimes cooperatively, sometimes disastrously, operating at different rhythms, scales, and temporalities. Only by embracing the fragility of our selves and of our worlds can we organize to advance an egalitarian, pluralist, and materially viable "new assemblage." The aspirational fascism substitutes the intensity of its unifying antagonism for the agonism of another affectivity: that expressed when the "thick ethos of pluralization dramatizes elements of unacknowledged suffering, uncertainty, and fragility in politics."[46]

For a political theology of the earth, the self-organization of the human collective that properly designates the political can no longer be abstracted from its *material*: from its enmeshment in its world of critically entangled, multiply human and nonhuman systems. The earth does not then appear as the unifying subject but rather as the ground, common indeed—a whole planetary underground of our lives. As systemic interactions with the nonhuman become more convulsive (too much water along coastlines, too little in prolonged droughts, too much heat and too little food), climate systems emerge in their own loops of self-organization. Global warming becomes its own system, as what Tim Morton labels a "hyperobject." It has become a "thing" that defeats our very notion of things.[47]

For our species, the critical difference encoded in the climate "thing" signals endless crisis, as the earth-ground refuses increasingly to play submissive background to the all-determining human polis. Earth's once trusty temporalities throughout human memory shaped the turning of the seasons of life, sometimes—as in the epiphany of "everywhere and forever" concluding Mahler's *Song of the Earth*—bursting into rapturous awareness.[48] These rhythms do not precisely desist. But they now push into the foreground modes of struggle, agonism, and indeterminacy that we had previously reserved for *human* history. No doubt they will both increase human ressentiment of our precarity, of temporality itself, while also stirring cannier responses to the contracting earth-time of our tense present.

Connolly pays special attention to the temporality, both the precarious finitude and the polyrhythmic complexity, of our cosmopolitical entanglements. Their recognition manifests the time—hear kairos—for a new political activism, indeed for a "democratic militancy" that will greatly exceed, though not omit, electoral politics. "The activism available today requires a large constellation of interinvolved minorities more than a core class surrounded by a series of fellow travelers."[49]

The multiple temporalities of diverse systems and their mutual imbrications accordingly suggest a new public strategy of "swarming." (We have something here to learn from the cooperative rhythms of the bees we are in the process of exterminating.) Gathering into a "pluralist assemblage" that emerges at multiple localities, its emergent commons buzzes with the planetary width of its variant undercommons. Its politics resembles what the *Cloud of the Impossible* frames for theology as the *complicatio* of entangled difference. "We inhabit," Connolly writes, "an entangled world in which the best hope is to extend and broaden our identities, interests, and ethos of interconnectedness as we multiply the sites of political action."[50] That extensive multiplication may in its intensive contractions produce—collect—the swarming political collective that organizes itself in the face of critical difference, difference first of all among its own diverse identities and priorities.

Surely, however, we must ask: How shall we "broaden our ethos of interconnectedness" at this time when time itself is contracting on us? We might, in the face of the entangled differences of earth crisis, ask instead: How can we not? If indeed "we owe each other everything . . ." The everything is variegated, planetarily out of all balance, as Cynthia Moe Lobeda shows of the concept of "climate debt" vis-à-vis race privilege.[51] Is our earthen interdependence—minded or ignored—itself a contraction, a *complicatio,* that folds difference not down but together? And if we would persist with a political theology of the earth, the contracting of humans in political self-organization must no longer abstract the polis from the bottomless buzz of its nonhuman components and companions.

Yet such an ecological undercommons does not suddenly homogenize—as though global warming now becomes our unifying foe—human diversity. That would just be a revision of the nature/culture binarism, the clean opposite of an ethos of interconnectedness. Critical differences that drive the unifying antagonism will not cease to press. But the pluralist agonism works them for complexity rather than mere contradiction. As chapter 2 will underscore, the agonism of a more common good can now never *not* involve the undercommons of nonhuman species, elements, systems. So its struggle plies the political as the process of complex self-organization: assemblage in the face of critical difference.

THE SOVEREIGN EXCEPTION

It is crucial to remember that the "critical difference" of political struggle demands the deep sense of *critical*—from the Greek, *krinein*, to "discern" or "judge," to "decide," linked to *krisis*, "crisis." Critical difference refers to a crisis in which it is necessary to decide upon action; this is why we assemble not just communally but self-organize politically. Critical difference, whether of perspective, identity, or situation, creates crisis and requires decision. And that can become dreadfully difficult, given the massed mess of our interconnections, resistant to the simplification of the cut, the decision—*de-cisere*.

Suddenly we find ourselves back under the laconic gaze of Schmitt, who has been waiting to quote to us the first major postulate of *Political Theology*: "Sovereign is he who decides on the exception."[52] The exception here means the emergency. Indeed, the power of the sovereign state is contracted into the person of the leader who decides, indeed who decides if there is a state of emergency—the lawgiver who may suspend the law. It is this exceptional power of the sovereign to decide, in what he [*sic*] decides is an exceptional crisis, that proves the rule of sovereignty itself.

It is right there, to the apex of power "in the exception," that Schmitt nails theology itself. If all modern political concepts are secularized theology, it is because historically "the omnipotent God became the omnipotent lawgiver."[53] And through this process of secularization law carries its hidden religious charge: "The exception in jurisprudence is analogous to the miracle in theology."[54] He then cunningly, disarmingly, paraphrases Kierkegaard: "The exception is more interesting than the rule. The rule proves nothing; the exception proves everything. . . . In the exception the power of real life breaks through the crust of a mechanism that has become torpid by repetition."[55] He mobilizes Kierkegaard thus against universalism—whether Hegelian, liberal, or Marxist. For the universal does designate an encompassing All, metaphysical or materialist, to which there can be no exception. Its rationality thus already constrains the power of sovereignty.

Schmitt fails to mention that, in the very passage in question, Kierkegaard goes on to write, in the irony of his own dialectic, that "the legitimate exception is reconciled in the universal."[56] For if the exception poses a true *novum*, then through great agonism—as of Abraham offering Isaac—it gets the attention of the universal. And this "legitimate exception" hardly finds the universal uninteresting: it "thinks the universal with intense passion."[57] Kierkegaard seeks to break a deadlock between the universal and the particular, not to torpedo the universal. And not only "liberal" theology defends a version of the universal. Thus for the sake of economic justice Kathryn Tanner classically plies a "theological vision of a univerally inclusive community of mutal benefit as our moral compass."[58] So in different language does a political theology accountable, at least, to the multiple communities of the earth.

Schmitt, who disdains any such encompassing good, is aiming at the torpidity of the liberal democratic bureaucracy, which blocks "the power of real life." And, unfortunately, the liberal machine does seem to yield the power to break through the crust of governmentality to the right. (Hence the successful power-grab of "real life," such as: "At the request of many, and even though I expect it to be a *very boring* two hours, I will be covering the *Democrat Debate* live on *twitter!*")[59]

For Schmitt, the bureaucracy of liberalism merges with the pluralist democracy of "everlasting conversation" (*ewiges Gespräch*), listening to all voices, practicing civility, and postponing decision. This modernist dissipation is for him inseparable from liberal Christianity or deism, "a theology and metaphysics that banishes the miracle from the world."[60] While liberal theologies had in the early twentieth century rarely questioned directly the dogma of divine omnipotence, they did downplay the imaginary of paternal control and intervention. So no wonder Schmitt, in the same passage, gets in a swipe at early twentieth-century experiments in reorganizing gender; he dumps them in the same category as the "anarchists . . . who see in the patriarchal family and in monogamy the actual state of sin."[61]

Too much hermeneutics, too much *her;* the secular potency of the sovereign Christian God was at stake. *His* omnipotence undermined, the state's capacity to act decisively in the time of emergency goes limp. That Schmitt eventually found his sovereign in the Führer reveals where the racial, sexual religiopolitics of antagonism trends—in the exception. The point is not that fascism flows inevitably from sovereignty, even from states of emergency or exception. No more is divine dictatorship the necessary consequence of any doctrine of omnipotence. But when crisis intensifies, authoritarian ascendency, breaking through the torpidity of flaccid systems, has its best chance.

NOTHINGNESS OR BECOMINGNESS

It is in a return to the figure of omnipotence as the divine source of Schmitt's legal doctrine of sovereignty that *Political Theology* culminates. Here he gestures toward something at first resembling the now-time of the kairos: to the "moment of decision, to a pure decision not based on reason and discussion and not justifying itself; that is, to an absolute decision created out of nothingness."[62] Anything else for him is lifeless universalism or else anarchy and chaos. Communist, socialist, anarchist, and liberal—these irreducible differences are fused thus into one galvanizing foe.

Decision "created out of nothingness" doubtless exercises an existential appeal; it ruptures the rationalized chronos, the compromising inertia, of modernity. As a political breakthrough, this "moment of decision" enacts the secularization of the theology of the *creatio ex nihilo*. The temporality of that doctrine of the creation must, however, not be confused with the time of the Pauline kairos. The ex nihilo signifies not the now-time but the single origin, as such, an absolute exception to time itself. The long Christian tradition of the doctrine of creation from pure nothingness does certainly interrupt any view of time as endless or mechanical continuum. But then there readily kicks in a presumption that this miraculous act presents the only alternative to the rationalized time of modernity—indeed, the only biblically warranted option. Yet, as most First Testamanet scholars have long agreed, the creatio ex nihilo is itself a postbiblical creation. I have shown elsewhere that the biblical narrative supports neither the omnipotent decision in the void nor the empty continuity of a cycle or a line of time.[63]

The *Elohim* of the narrative that opens the Bible can be said to *decide* in favor a world. But that decision is emphatically not "out of nothingness." It is not "out of" nothing as mere, pure void, but of the complex potentiality of what the second verse calls "the deep." That *tehom*, translatable as "chaos" or "ocean," is what Augustine rendered as the "nothingsomething."⁶⁴ Yet of course no Schmittian political theology can admit a hermeneutics of creation from such oceanic chaos, that dark and indeterminate, indelibly Jewish, tehom. Indeed, no straight story of creatio ex nihilo, whatever its politics, will tolerate the counterdoctrine I dubbed *creatio ex profundis*. It floods the foundations of the whole doctrinal edifice of God's omnipotent control. Through its lenses, the narrative of Genesis itself begins to read as an "everlasting conversation." Indeed the eleventh-century commentator Rashi finds in the plural of "let us create" evidence that creation is a consultative process.⁶⁵ The plurisingular Elohim calls,⁶⁶ the becoming creatures respond in creativity, and Elohim answers back, in pleasure, "it is good," luring a whole ocean and planet into the evolutionary cocreativity of a self-organizing cosmos. The particular cosmos that emerges appears as altogether new and good, good, good—but not as an "absolute decision" in a void.

Operating underground, or underwater, as a political theology, the "original" creatio ex profundis might join again with the contracting kairos, in which there takes place always again a new, open-ended, and therefore continued, creation. In taking place it is taking time. In its moments of evolutionary breakthrough, in its kairoi of novel emergence, the creativity of *genesis*—the Septuagint Greek for "becoming"—delivers not the unilateral rule of omnipotence but the Spirit that "hovers over the waters," not the absolute decision but the call, "Let there be . . ."

For political theology this hermeneutical nuance has immense consequences. The theology of the creation from the deep would pump another secularization than that of the sovereign exception. Let us call it the *inception*. For the chance of the new, its real possibility, urges us to begin again, at a present moment of crisis, a present confrontation of critical difference. And again, and . . . Unlike the movie *Inception*, this novum does not exercise control over another's unconscious mind. But, like it, a radical change of practice is intentionally initiated, its theology seeking subliminal insertion into secular politics. It calls to us not in a void but in the chaos of our genesis, in the messes—which may feel anything but messianic—of our becoming.

In the becomingness of the inception flashes the resistance to each self-declared exception. Freed of any single, sovereign Decider, political theology *divines* another possibility: the multifaceted public embrace of planetary entanglement.

ONE EMPIRE, ONE RELIGION

Political responsibility—for theists, atheists, naturalists, and others—therefore takes account of new operations of old schemas, which is to say, attention to the ongoing political operation of divergent, often opposing, theological motifs. Attention is not conversion. It does require some recognition of the agonisms within and between theological traditions and so between their conflicting secularizations. In other words, as "political theology," that ongoing operation marks a methodological back door by which living religious thought, even in specific forms (Indigenous, Hindu, Buddhist, Jewish, Christian, Muslim, etc.) today may enter a livelier breadth of conversation. That broad discourse stays vibrant to the extent that it prepares the way of a multiply common good.

If that pluralist culture, as always already broader than the secular, is sometimes called postsecularist, there is meant no simple "return" to or of the religious.[67] It is not reducible to surges of fundamentalism or dreams of a Christian America. The recognizable resurgence of religion in multiple progressive movements (as in a half-century of liberation theologies) represents neither a regression from nor a progression beyond "the secular." So the reactionary option does not represent mere desecularization, any more than the progressive represents mere secularization. For what is at stake is not secularity as such but the intensification of some as against other schemas of secularization.[68]

Schmitt was attracted to the sanctified monarchicalism of theocracy. Such nostalgia helps clarify the conservative privileging of the legacy of exceptionalist omnipotence even or preicsely *in its secularization*. The model of monarchical sovereignty was never the only political theology of which Christianity was capable. But that schema's justification is recognizable in overt form at least as early as the fourth century: "And surely," writes Bishop Eusebius, Constantine's house theologian, "monarchy far transcends every other constitution and form of government; for that democratic equality of power, which is its opposite, may rather be descried as anarchy and disorder."[69] I confess to being startled by such overt antidemocracy at the source of Christian empire. His political theology is utterly self-consistent: "And thus by the express appointment of the same God, two roots of blessing, the Roman empire and the doctrine of Christian piety, sprang up together for the benefit of men." Eusebius mobilizes the unifying force of monotheism against *all* multiplicity and equality. And so the One Empire is the twin of the One Religion.

This commitment to the mon-arche, the single origin with its all-unifying rule, did set him in uncomfortable tension with the symbol of the triune equality of the "persons" that was crystallizing at that very time of Nicene creedal consolidation. (His defender Schmitt was similarly, despite his interest in reactionary movements of Catholicism, in tension with trinitarian theology.) The Byzantine *pantocrator*—the Son as the all-ruler, enthroned just above the throne of the Christian caesar—presided over the consolidation of his Father's properties on earth. The internal egalitarianism of Persons could then be maintained as itself the exception to the hierarchy of external relations.

So we witness unfurling from the Constantinian *saeculum* a long history of the *theo-logic of the sovereign exception*. Enthroned as the rule, it yields a potent historical chain of political exceptionalisms: first of all, that of the imperial Christian supremacism; of its corollary religious exclusivism (*extra ecclesiam nulla salvus*); and of the dehistoricized and de-Judaized Christ who rules precisely by supernatural exception rather than by human example. As we shall see in the following chapter, that sovereign christology powers, at the same time, the dominion of Man as the anthropic exception that would drive in secularized form the whole modern nature-conquering project. Now, however, let us we observe how this political theology of the enthroned exception plants itself right in the antimonarchical and secularizing origin of American democracy.

WHITENING EXCEPTIONALISMS

Here I depend upon the genealogy of American exceptionalism offered by womanist theologian Kelly Brown Douglas in her all too timely *Stand Your Ground: Black Bodies and the Justice of God*. "In an effort to establish the antiquity of the Church of England," this Episcopal priest writes, "Archbishop Matthew Parker (1504–1575) encouraged research into the . . . history and politics of Anglo-Saxons."[70] For two centuries the research centered in a reading of Tacitus' *Germania* (98 CE), an ancient Roman ethnography of the blue-eyed tribes which the English claimed as their ancestors. This research into ethnicity was driven by an anti-Roman—that is, antipapist—sentiment. At the same time, the study also nourished participatory governance: "According to Tacitus, within the various tribes 'the whole tribe' deliberated upon all important matters, and most final 'decisions' rest with the people."[71] The description of this proto-democratic self-organization fed such radical political experiments as the

Levellers, the Puritans, the Pilgrims. The Anglo-Saxon myth came to America through the English reformers, the new Israelites in God's plan: "They considered themselves the Anglo-Saxon remnant that was continuing a divine mission . . . traced . . . beyond the woods of Germany to the Bible."[72]

The remembered moment of the Exodus from slavery stirred—as it did in all the radical movements of Europe—the kairos of liberation.[73] Political sovereignty began its deep shift from the monarch to "the people." There took place new experiments in self-organization across an ocean of critical difference. Its messianic inception cannot be divorced from its liberative history, nor reduced to a single motif or color. The theopolitics of the New Polis had its moment. Even that notoriously exceptionalist city shining on the hill had originally glowed with emancipatory energy. It alludes to a parable of Jesus, who with characteristic irony says to his motley homeless band: "You are the light of the world. A city that is set on a hill cannot be hidden" (Luke 5:14).

Its light however can quickly dim. Once having settled along the coast of the New World, the Anglo Saxon exceptionalism begins to shift toward an overt and secularized racism. None other than Benjamin Franklin would observe that

> all peoples are black or tawny, even the Germans, the Saxons only excepted, who with the English make the principle body of white people on the face of the earth. I could wish their numbers were increased. And while we are . . . scouring our planet by clearing America of woods, and so making this side of our globe reflect a brighter light to the eyes of inhabitants in Mars or Venus, why should we in the sight of superior beings, darken its people? Why increase the sons of Africa, by planting them in America?[74]

Yes, fellow white Northerners, in this letter's bit of extraplanetary whimsy, Franklin delivers a racist logic for opposing the transatlantic slave trade. These lines express an early and private musing of Franklin, but we glimpse herein the fond vision of a shining nation cleansed at once of dark woods and dark people. For a political theology of the earth, these twin images indeed belong together: the white and the anthropic exceptionalisms here fuse in one take-out dream.

Note, by the way, that "exception" comes from *excipere*, the Latin "to take out." It works both ways: the sovereign exception takes itself out of the common; then the exception can "take out" whatever impedes its ascent. So the takeout of the letter is the double-edged vision of American progress toward a planetary ecoracial sovereignty. We are still taking out the human and the nonhuman obstructions to our bright white supremacy. And in the process we may be taking ourselves out—of the web of planetary life.

The particulars of the story would continue to shift dynamically. But, as Brown Douglas deftly demonstrates, "the narrative of Anglo-Saxon exceptionalism is America's exceptionalism."[75] The challenge to national unity soon became the multiplicity of immigrant stocks. That diversity was felt as emergency by the late nineteenth century: "President Theodore Roosevelt ... feared 'race suicide,'" and, Brown Douglas shows, it was the construction of "whiteness" that worked "to resolve the contradiction between America's Anglo-Saxon and immigrant identity. Whiteness signified that the immigrants were Anglo-Saxon enough."[76]

And therefore—human enough. For whiteness at the same time had come to characterize the normative separation of the human from the nonhuman, from the animal, from "nature." At this juncture, the philosopher of religion Carol Wayne White's elegant articulation of a "black naturalism" calls us to the difficult, simultaneous confrontation of racism and speciesism: a yoked pair of exceptionalisms crucial to the modern sovereignties of empire and economics. She extends her naturalist ethics, grounded pointedly in the "sacred" and not in the "supernatural," into the entire cosmic "matrix of complex interconnectedness."[77] Perhaps only such a matrix—close to Connolly's "ethos of interconnectedness"—can surround and sabotage the scheme of differential separations sustaining the hidden hierarchies of the sovereign exception.

MESSIANIC WHITE MAN

Let us pause to take *in* how a versatile religiopolitics of race fuels the secularized galvanizations of a we versus a they—a we to be unified in being cleansed of the dark taint of multiplicity, a they ever subject to being "taken out"—in police violence, in immigrant deportation. In its most forthright antagonism, constitutive of United States history and therefore saturated with resentment, this politics recharges the decisive clarity of the color line, of white versus black. At the same time, the complexity of the multiple—of shades of unwhite complexion, African or Latin American, Asian or Muslim, of tones of sex or gender—operates in great asymmetrical histories of entangled difference.

Sometimes waves of democratizing agonism work the edges of chaos toward new self-organization. But at crisis points the sovereign whiteness curdles again into lumps of affective oneness. The racial "they" becomes the direct object of antagonism when *they* name and resist their subjection, threatening the brittle "we" with specters of insubordination, intermarriage, or insurrection. And so they can be cast as convenient aliens for the politics of the foe, making "friends" across vast class and cultural difference between whites.

Race as definitive of ethnic difference belongs to a quite modern schematism of human self-organization. Slavery goes back to the origins of civilization, at least to the building of cities. It was as likely to be justified by victory in war as by ethnicity; and ethnicity was more categorized by language and other cultural differentiators than by a unifying biology. In *Race, a Theological Account* (2008), J. Kameron Carter makes a persuasive argument that it is the late-medieval European racialization of "the Jew"—previously a "people" or an *ethnos*—that reorganizes the human in terms of the notion of race. The race machine is then from the start a modernization (but not immediately a secularization) of a Christian supremacism that is inseparable from Christian supercessionism.

So we witness a Christological exceptionalism that takes something called Christianity out of its Judaism, takes someone called Jesus out of his Jewish body, takes something called *christos* out of its messianicity.

Note also that, on the Schmittian model, the exception works at once to take the sovereign out of the law, as *above* it, and to take the religio-racial other out of the "we"—and so out of the protection of the law—as *beneath* it. More recently, Carter has offered an analysis of race precisely in terms of political theology. He exposes how the constitution of "the White Masculine as imperial Man was tied to his assuming a messianic and mediatory role in the world." Thus Carter captures (in advance of its garish recent manifestation) the shining toxicity of a hetero-male supremacist religiopolitics: as "imperial Savior, he [is] sanctioned by divine right."[78] In other words, the secularization of a theological schematism of omnipotence that invests divine sovereignty in the ruler, the nation, or the people then finds analogous deployment in the modern secularization of the Lord as White Man.

The problem lies then first of all not with the messianic—distinctly notwhite in its biblical materializations—nor with the messianic and secularizing intensifications of the struggles of any oppressed people. Vast pyramids of structural injustice precede the biblical exodus, the prophetic self-critique of Israel, and the hope of a messianic liberation from external and internal suppression. The white pyramids of modern racism are built with an insidious Euro-American absorption of the energies of Exodus and its legacy of messianic eschatology. Whiteness in the United States had around it enough colors of the human to deem itself the exception as the threatened majority. So if the appropriation of messianism by way of a Christian supremacism always fuels the political theology of the sovereign exception, it is from the U.S. version that shoot the ongoing, versatile infusions of White Manhood. One might therefore reasonably give up on the messianic altogether—what's the difference between messiahs, after all? But a political theology mindful of itself trusts more to differentiation than to erasure.

CHARISMA AND THE WOLF

It was Giorgio Agamben's reading of Schmitt that provoked his return to the apostle Paul's figure of the messiah, the christos. Agamben decodes Paul by way of such leading continental Jewish thinkers of the Second World War era as Walter Benjamin and Jacob Taubes, with their own complicated engagements of both Schmitt and Paul.

Agamben starkly distinguishes Paul's sublation of the law as a fulfillment, "the messianic *pleroma* of the law," from the state of exception which indefinitely suspends the law.[79] In *Homo Sacer* (1998), Agamben had analyzed the reduction to "bare life" of persons or of populations excepted from the human condition. He has been long preoccupied with the analogy between multiplying European refugee camps for immigrants in the late twentieth century and the Nazi *Lager* (U.S. detention centers for immigrant families were yet to come). He was then provoked by the U.S. Patriot Act of 2001 to write *State of Exception* (2003). The state of exception "not only appears increasingly as a technique of government rather than an exceptional measure, but it also let its own nature as the constitutive paradigm of the juridical order come to light."⁸⁰ Reflecting on the routine U.S. violations of international law regarding torture and imprisonment, he concludes that "the state of exception has today reached its maximum worldwide deployment."⁸¹ That may sound hyperbolic for 2003. But the U.S. state of exception continues to obtain (e.g., in Guantanamo Bay). And now we may hear new resonances: "The normative aspect of law can thus be obliterated and contradicted by a governmental violence that—while ignoring international law externally and producing a permanent state of exception internally—nevertheless still claims to be applying the law."⁸²

Agamben is here summoning Walter Benjamin, who wrote of fascism that the "state of exception had become the rule."⁸³ Contemplating Schmitt's figure of rule by suspension of the rules, Agamben warns against the collapse of two distinct powers into each other: *potestas*, as "normative and juridical," and *auctoritas*, as anomic and metajuridical, an authority emanating from the personality of the leader. When they fuse in one person, "the juridico-political system transforms itself into a killing machine." To "try to interrupt the working of the machine that is leading the West toward global civil war" requires that we separate these forces.⁸⁴

More recently, in a *Lawfare* blog post just after the 2016 election, legal theorist Quinta Jurecic cuts to the quick. With reference to Agamben, she analyzes "the nightmares of those who thought they saw Schmitt in the form of George W. Bush. Certainly the Bush administration saw an intentional turn toward an aggressive view of executive power."⁸⁵ But as she points out, though pressured to the point of a Supreme Court decision, "the fundamental structure of the rule of law itself has remained standing." So did Agamben cry wolf? "Our new president-elect . . . now poses an interesting problem for the Schmittian revival: have we now, eight years after Bush left office, elected our first Schmittian President?" With a closing flourish she reminds us of the crucial bit of Aesop's fable: "The wolf actually does show up at the end. So, is Trump the Schmittian wolf that Bush was not?"

With apologies to real wolves, the fable's sheep would have the answer to Jurecic's question. And setting aside the question of whether 45 had any knowledge of a state of exception, indeed of international or constitutional law beyond the racialized, sexualized potestas of "law-and-order," he was always characterized, whether in admiration or dismay, as one who does not play by the rules. And a critical mass felt from him an exciting affect of exceptional authority.

Agamben's auctoritas develops Max Weber's concept of the "charismatic leader." Weber had contrasted both the religious and the political charismatic to the other styles of leadership, which he dubbed "traditional" (conservative) and "legal" (administrative/bureaucratic). Charisma signifies "a certain quality of an individual personality, by virtue of which he is set apart from ordinary men and treated as endowed with supernatural, superhuman, or at least specifically *exceptional* powers or qualities."[86]

For Weber, charisma is in itself ethically neutral. The religious aura of *charisma*, from the Greek for "grace," never quite dissipates. Its theological background is manifest in the originators of religious movements. He offers the (now so timely) example of Saint Francis, whose charisma was felt by nonhumans and humans alike. In terms of the temporalities of sovereignty, we might then infer that traditional leadership repeats a normative past, whereas democratic or socialist legal institutions presume a gradualism of progress. The charismatic however would seem to lead from an intense present tense—a now-time that, for those feeling its forcefield, ruptures the homogeneity of chronos.

So the religious aura of the *charis* glistens through its secularizations. These may manifest in the kairos-moment—indiscernibly secular/religious—of a Gandhi or a King. Or, on quite the other hand, the "exceptional powers" may pump the fusion of auctoritas and potestas that yields the popular strongman and his killing machine. On the one hand irrupts a charisma of public love; on the other, ripping free of compassion, the sovereign charisma of the bully. The former conveys the messianic potential of a new earth and its public; the latter, the messianic farce of White Man. Two different theologies, two different secularizations.

The messianic charisma cannot, it seems historically, successfully compete with the potestas of its antagonists. Even in outrage it does not match the antagonism. Yet its practice is not the pacifying acquiescence often confused with just love and forgiveness. It yields what in this theological context we may call an *amorous agonism*: the forthrightly theological vocalization of Mouffe's and Connolly's respectful agonisms. It lacks potestas, but it claims authority. It is gifted with what, according to Benjamin (on the verge of death in his attempt to escape fascism) every generation is endowed with: "a weak Messianic power, a power to which the past has a claim."

That past claims the gaze of his "Angel of History," in the famous allusion to his sketch by Paul Klee, *Angelus Novus*—an icon of an icon. Benjamin read the drawing thus: "An angel looking as though he is about to move away from something he is fixedly contemplating.... His face is turned toward the past. Where we perceive a chain of events, he sees one single catastrophe."[87] He reads the angel contemplating in horror the unredeemed losses, the collective traumas: the lynchings, the camps, the bombings. "But a storm is blowing from Paradise..." It "irresistibly propels him into the future to which his back is turned."[88] The backwards gaze of a weak power, angelic or messianic, theological and political, does not predict or direct the future. The messenger, *angelos*, presages no omnipotent redemption. But the angelic compassion is not therefore reduced to that indissociable past, does not lack its own *potentia*.

Indeed the amorous agonism does not cease to blow or break into renewed political potency, into movement theologically inflected. A later carrier of its message: "Love is the only force capable of transforming an enemy into a friend." Capable, and without guarantee. Such messianic agape as Martin Luther King's, even while not left to the angels, does not take the place of politics. But neither can this love be captured in apolitical privacy. With its history of effects, limited but indelible, it refutes the presupposed antagonism of Schmitt's political theology.

SENSUOUS INTERSECTIONALITIES

At a later time, the angel's gaze is caught by the dying lives of sea and land wrapped in a haze of what could have been otherwise. If then amidst chaos the messianically weak power inaugurates a future, it sketches the barest beginning. What may appear first as the exception has then swerved into the inception. Its singularity, almost as though from nothing, reveals its novelty in fact entangled in a dense precarity of relations. So, in that swerve, this beginning does not abstract its future from its past (a charismatic temptation, no doubt) but contracts instead in its kairos the entirety of its history. Nor does the passion—the "being moved," the lament—sever the struggle from the desire for the yet-to-come, the *eros* that inflames an amorous excess.[89]

If that weak power has also been encrypted as crucifixion (not in Benjamin's writings) it is because, despite its later triumphalism, the cross endlessly recalls an agony that cannot be erased. In fact the cross performs politically a double coding: of both the sovereign antagonism and the amorous agonism, of the opposed world schemas intersecting. The former nails above the suffering body of Jesus a sign, *The King of the Jews*, in sovereign ridicule of the messianic promise; the latter inspires a long history of struggles for the *basileia theou*, the kingdom of the least, the parody of power. The sneer came later, when imperial force posed as the Church's truth and victory trumped the agonies of struggle.

We can distinguish here between the ongoing effects of two histories, two great currents of secularization, two political theologies: that of the omnipotent sovereignty in the exception and that of a messianic commons in the inception. One might then read against the counterrevolutionary Schmitt his contemporary, the Marxist, atheist, and Jewish Ernst Bloch: "Messianism is the red secret of every revolutionary."[90] Bloch is the philosopher of the "not-yet-being." He excavated in rich genealogical detail the history of biblical eschatologies and "Christian social utopias" as the source of all Western revolutions, democratic and socialist—indeed as the wellspring of what is called hope. But then "true hope moves in the world, via the world," expressing a "not yet," he wrote, "which in the core of things drives toward itself, which awaits its genesis in the tendency-latency of process."[91]

Bloch's strong influence on José Esteban Muñoz in *Cruising Utopia* (2009) illumines one recent road to the secularized messianic commons. In a famous exchange, Muñoz respectfully rebuts Lee Edelman's brilliantly pessimistic "no future." If Edelman demolishes political hope as the issue of the reproductive time line of heteronormative futurity, Muñoz answers with the political queering of hope. "As strongly as I reject reproductive futurity, I nonetheless refuse to give up on concepts such as politics, hope, and a future that is not kid stuff."[92] Speaking for "queers of color," Muñoz enacts a "not-yet-here" of queer performativity for the sake of the "political imagination." We need his indelible language:

> It is important not to hand over futurity to normative white reproductive futurity. That dominant mode of futurity is indeed "winning," but that is all the more reason to call on a utopian political imagination that will enable us to glimpse another time and place:

a "not-yet" where queer youths of color actually get to grow up. Utopian and willfully idealistic practices of thought are in order if we are to resist the perils of heteronormative pragmatism and Anglo-normative pessimism.[93]

So Bloch's messianically derived "not-yet" reaches back through and beyond class into the aching entanglement of race and sexuality. But we note how the thrust of his argument resists even a same-sex exceptionalism: "Imagining a queer subject who is abstracted from the sensuous intersectionalities that mark our experience is an ineffectual way out. Such an escape via singularity is a ticket whose price most cannot afford."[94] Refusing the ticket "out," it may be said to perform an alternative also to what Jasbir Puar calls "sexual exceptionalism," in which a certain elite homosexuality has also been rendered complicit in a white supremacist, particularly Islamophobic, politics.[95] Refusing the escapist singularity, the "sensuous intersectionalities" reverberate with the possibility of a collective inception.

Without a queer multiplicity of love strategies resisting the straight-white-manhood of a messianized *potestas*, what potentiality for an amorous agonism can there be? As politically theological, love either locks into the merely private—or it heats up the struggle against every sovereign supremacism.

HAZARDOUS HOPE

Muñoz does not discuss Bloch's theological derivation of all political hope. Yet, at a different tilt, Bloch's influence on Jürgen Moltmann and his early classic, *Theology of Hope* (1964), opened an originative pathway for Christian political theology. Relinking Christian thought to the ancient prophetic practice of dissident eschatology, Moltmann tugged hope for the first time to theological center stage. It appears now not as passive expectation: not a predictive knowledge, but as "the passion for the possible."[96] Moltmann has, in the evolution of his own confessionally Christian vocabulary, never ceased to think with Bloch's revolutionary "thinking of hope." But Moltmann would insist that "without hope for the ultimate, hope for the penultimate soon loses its force, or it becomes violent in order to extort the ultimate from what is penultimate."[97] In other words, hope readily forms idols of its ideal future. Then certainty takes the place of faith. And revolutionary antagonism can almost compete with its reactionary opposite in historic body count.[98] So if political theology *minds* its theology—as discourse of the ultimate—it resists all totalizing hope for history. But not all revolution.

Of course, any messianically charged hope, avowed or not, faces in its secularization the temptation of a modernist faith in dialectical progress. A radical demand for justice opens it to faith in a promised outcome. And so with most theological eschatology the promise morphs into the guarantee. (Indeed in *The Theology of Hope* these appear indistinguishable.) If the revolution drives not just with determination but with determinism toward the utopic end—knowable to its radical elite—it moves from total certainty to totalitarianism. And it moves *straight*. But any progressive desire for a world of justice will be haunted by the might-have-been of socialism and therefore also its might-yet-become. To heed what Derrida calls the "specters of Marx," the haunting voices of the biblical hope for justice, charged with "messianicity" rather than "messianism," is to hold the politics responsible for its theology—and so vice versa. And the very pastness of the spectral present keeps the present possibility of the not-yet in play. That backward gaze keeps hope from abstracting itself from its sensuous now.

Utopia remains what it literally means: no place. The to-come may not come; it does not exist somewhere just waiting to perform the apocalyptic exception to all prior history. And yet it names something real if not actual now: the possibility not yet actualized but *presenting*. The inception signifies that present possibility—as active potentiality, which is to say, as virtual reality. The inception lives in the indeterminacy of now. It rumbles through the undercommons, it intersects multiply, it springs from the earth-ground of its time. So it can be said to recycle its own history, to contract and compost it as the place of what Whitehead calls "real potentiality." His teaching of the togetherness of becoming influenced Bloch's "genesis" awaited "in the core of things," in "the latency-tendency of process."[99]

The past as present potentiality can poison the loving struggle of the vulnerable now. Or nurture it. The work of hope, far from any sanguine certainty, is what Bloch calls a "hazardous business."[100] In its public urgency, it does not postpone or predict the not-yet so much as it begins to perform it—to ingather, to contract it. So then hope cuts free of the numbing cycle of optimism and pessimism. Optimism and pessimism both *know* the outcome. The hazardous hope however does not know, does not predict. Nor does it patiently await. "Draped in black," this hope grieves even as activates. Even now.

KAIROS AND INCEPTION

The mottled history of collective struggles for justice, divergently secularized in breakthrough moments of resistance, revolution, *satyagraha*, civil rights, Occupy, Black Lives Matter, #MeToo, climate marches, high school walkouts against gun violence . . . This variegated sea of social movements delivers its specters to our present. Its moments of messianic collecting pose an alternative political history, with its own charismatic now-times, to that of the sovereign economico-political powers.

The inception even in its revolutionary potential does not, however, define itself in antagonism to sovereignty as such. The concept of sovereignty remains sometimes indispensable to the struggles of the global South against the uber-sovereignties of U.S. and global capitalist dominance. Sovereignty may in certain contexts resist its own temptation to the exception and strengthen democratic, socialist, and indeed ecological alternatives.[101] The alternative to the sovereign exception is not one. Its inceptional assemblage contracts and collects ever again in the struggle with critical difference. The crises of difference inside and beyond its own gathering for a more common good do not cease.

If the sovereignty of the exception seemed recently to pull off a great American fusion of potestas and auctoritas, its persona—an unhinged Caesar wannabe with nukes—lacks the dignity of the Latin. In its power grab it violates the precarious becoming of democracy, vulnerable even as it violates the fragile commons of planetary life. At this moment, hope does strain against the impossible. It runs out of energy, out of fashion, just out . . . Toxins of racialized, sexualized, financialized, americanized, carbondioxidized power are penetrating our common world, our collapsing schema.

Time is contracting. May we remind each other that what confronts us now as impossible might crack like Leonard Cohen's broken hallelujah into unforeseen possibilities? In a moment. That might be the moment of the kairos, the messianic now-time that Agamben nimbly traces, following Taubes, through Benjamin's hidden Paulinism: "History," writes the latter, "is the subject of a structure whose site is not homogenous, empty time, but time filled by the presence of the now (*Jetztzeit*)."[102] As an unorthodox Marxist, Benjamin found that the schematic modern homogeneity of time had been absorbed into a Marxist vision of inevitable progress and so had betrayed the working-class struggle in Germany. He opposes to modern temporality the *Jetztzeit* that through Agamben translates Paul's *ho nun kairos*.

"The Jews were prohibited from investigating the future," which is to say, from the predictions and certitudes often confused with prophecy. Yet neither was "the future turned into homogenous, empty time." Benjamin lacks all optimism. But one can hardly call the mesmerism of the last sentence of his *Beiträge* hopeless: "For every second of time was the narrow gate through which the Messiah might enter." The gateway of every now iterates the temporal schema he had jut encapsulated as "a model of Messianic time." It "comprises the entire history of humanity as an enormous abridgement" or, more precisely, "monstrous abbreviation," *abbrevatur*.[103] The recapitulation thus contracts an immense history into itself. Into its kairos.

Might we hear the echo of the Corinthian sunestalemnos, the contraction of time itself? Imagine the pressure of that crystallization. You might feel it now, as you gather this historical moment, in its agonies, its fragilities, its possibilities, into yourself, as you contract in present tense the personal-planetary history that makes you up. The tensions, the contradictions, that pressure this moment do not shorten time so much as they squeeze intractable difference, opposing demands, into it—including the binaries that form a we against an indubitable enemy (a figure, a party, a policy, a class, a global economic schema). Yet the very contraction of the moment complicates—folds together—and therefore, if we let it, strengthens the democratically and socially justice seeking public.

Yet to assemble across critical difference invariably confronts us with chaos, a messiness that precisely crosses the boundary of us/them. And so of me/you. It requires confrontation also of internal contradictions (my collusions in race, class, and straight-married securities as one unexceptional complication). We do not wait to achieve purity. For that time *is* too short. But we might work mindfully with conflicts, up close and planetary. If creation is from chaos (never mere disorder), might the very mess of our history press open the messianic gate?

In the Schmittian view, however, history becomes civilization only in the imposition of order, which is to say, in the *deferral* of the messianic coming. So he reads Paul's *catechon*, the restrainer (2 Phil. 2): "The belief in an arresting force that can stave off the end of the world is the only link leading from the eschatological paralysis of every human action to such a great historical agency as that of the Christian empire at the time of the Germanic Kings."[104] So the postponement of the end thus sanctifies imperial sovereignty, in all its contradiction of the messianic hope. But the reference to the catechon in 2 Philippians, a letter most scholars consider to be pseudo-Pauline, is nowhere repeated in Paul. 1 Corinthians teaches fearlessness—and so resistance of socially normative self-definitions (like marriage)—precisely in embrace of the kairos. Taubes recognizes Schmitt's "catechonic impulse" as the counterrevolutionary drive to save "the state from the chaotic powers of the party," containing the unruly masses as long as possible. With characteristic wit, Taubes notes that Christianity has also practiced this catechontic sovereignty: "one prays for the preservation of the state, since if God forbid, it doesn't remain, then chaos breaks loose, or, even worse, the Kingdom of God." To prevent that revolutionary outcome, one must "capture the chaos in forms, so that chaos doesn't take over."[105] Yet it should not be misread as a mere preservation of order. For it is always some form of chaos that justifies the state of emergency.

Emergency, whispers Angelus Novus, need not serve the sovereign state of exception; it can trigger instead the inception, the emergence of a new collective. Its real potentiality was, for an exemplary moment, experienced by millions around the earth in the January 21, 2017, women's marches. Something fierce, fresh, promising, splicing rage with humor. And since that moment, what charisms of self-organization at the edge of chaos have arisen? We could collect ourselves yet. Haunted by that agonizing, tantalizing kindom, the specter of a new public intersects sensuously with a multitude of mattering collectives.[106]

TENSIONS OF PRACTICE, FOLDS OF CONTRACTION

Yet even among the invitational "we" of those likely to join such movements and marches, critical differences flare; decisions, with their (sharply) cutting edges, cannot be eluded. The nowtime in its contraction is rent with contradiction within the very self-organizing process of its complex assemblage. So then consider four common binary tensions, prone but not doomed to antagonism, of present practice.

Social Movement/Electoral Politics

There will be no political transformation for a more common good unless we build upon the social movements growing as cells anywhere, networking virtually everywhere, and spilling as crowds into the streets. Refusing politics as usual, do we cease to engage in electoral politics? We would then share Schmitt's right-wing view, mirrored by much of the left, of liberal torpidity. But how should we trust the historic capacity of democracy to transcend its sellouts?

So then, not in trust of any progress but by insistence on just process, might we strengthen, radicalize, even socialize democracy? Might we support some who assemble with us as they evince charisms for public office and stomach for institutional struggle in its impurities? And, at the same time, might we heed the inseparability of inception from insurrection, speaking nonviolently, indeed theologically? Thus Clayton Crockett:

"Insurrectionist theology affirms that everything must be rethought, everything is unsettled, even the Earth we stand on and take for solid foundation."[107]

Issue/Coalition

How do we intensify the struggle on behalf of particular, and particularly vulnerable, populations? Black Lives Matter, reproductive freedom, sanctuary for immigrants, LGBTIQ rights, #MeToo, student survivors organizing for gun control? What about class and its economics? And the climate of them all? Do we choose one issue and double down? At what point does that mean that we unify around our single issue identity—and so our own progressive exceptionalim? If, on the other hand, we work for the broad intersectional coalition that might multiply and, well, *work*—does that mean to settle on too low a common denominator? Or maybe not low enough. What if we go lower yet, get common enough for an uncommon solidarity with the undercommons? Beneath foundations, where the intensity of my issue rocks into the movement of a multitude, issuing not in unity but in an immense coalescence?

Locality/Globality

And yet "the multifaceted image of pluralism" advanced in modernity can "sit rather comfortably with this or that creed of human exceptionalism."[108] But the living contexts, the actual localities where new human assemblages collect, now refuses to hide its underground radicality: its radix, roots, unfurling in the earth. But that inescapable commons of our materiality has been schematized as the globe. The globe, unlike the planet, may threaten to smooth away every locality, and so every particularity, to trap it online, to make it a deal. Thus Gayatri Spivak had proposed "the planet to overwrite the globe." In postcolonial analysis of neocolonial economics, "globalization is the imposition of the same system of exchange everywhere. The globe is on our computers. No one lives there."[109] Thus, according to the political ecotheology of Michael Northcott, "The idolatrous universal—monetary accumulation—commits corporations, financiers, and governments to the destruction of places both far and near." Thus we are seeing populist reactions against liberal and neoliberal universalism successfully manipulated by the political right.[110] Local place vs. global city? Yet the political comes entangled in the polis, its theology in the urban Paul. How can it simultaneously reroot us mindfully in place, *in time*, while exposing the idolatrous *local*—of resentful identity, of ethnic belonging, of excepting itself from its unsettling interdependence with stranger places?

In the vibrancy of our local alliances, might we find the strength, the spirit, and the numbers of an earth that flows through us materially at—virtually—any place and time? Might ecosocial self-organization not set loose new schemes, mindfully glocal, dauntlessly planetary, for survival no matter what? That planet will step up as the subject of chapter 2.

Religion/Saeculum

As to political theology: given the growing secularism of the public, of progressive politics, let alone of the Democratic Party, should we maybe just give it up? If political theology is about secularization, perhaps it should just secularize itself once and for all? Cut free of the endless hermeneutical perplexities of these old patriarchal authorities? After all, the religious left already knows how to exercise the coalitional rigor of secular language.

But even utilizing that translational capacity, politically indispensable icons of messianic charisma—Gandhi, Martin, Malcolm, let alone either foot-washing Francis—cannot be scrubbed clean of their theologies. And movements within the world religions for gender, racial, sexual, economic, and ecological justice keep happening, all pluralist in their ecumenism, all steeped in secularity. Given the theological language propelling fundamentalism and terror on the right wings of the Abrahamic traditions, we might still question whether a fundamentalist secularism is the answer. Also secularism (again, not secularity) culturally aligns the North against the global South. So how does its antireligious version not serve white neocolonial globalism, particularly in its Islamophobic status quo?[111]

The conundrum of theology as such exposes then the following assumption of the present experiment: even the most secular versions of political theology expose the impossibility of excepting theology from politics. Or then the secular from the religious. "*We are secular-religious Others.*"[112] Perhaps a theology that counters its own sovereignty of sovereignties can only now, in the time of a weakened Christendom, come into its own. Might theology—not just as the ghost of a past or the fiction of a future but as the *kairotic discourse of the presenting possible*—help now to gather a spreading public of shared earth struggle? Even or especially amidst the contraction, here and there rapid, of the religions themselves? If so, it will minister to each of the aforementioned binaries, stimulating a more respectful, indeed maybe even a more amorous, agonism. Then the tensive differences may mark the folds of a contraction, not the cuts of an exclusion.

When confronted with the crises of collecting ourselves as a new public, we do not answer such internal tensions with a simple decision, one side or the other. Such decisionism forfeits for us the intersectional potential offered in the monstrous contraction of the political present. That potential may be presenting itself in versions of the *coincidentia oppositorum*—the coinciding of opposites expressed by an old negative theology—that give us pause and then, in the gateway of that pause, open a third way. What if we let the amorous agonism spiral back through all our most pressing political questions, collecting us not in purity and oneness but in entangled difference? Decisions then are not avoided but deepened. They take place not out of nothingness but amidst the chaos.

No longer surrendering to the sibling rivalries of the academic left or to our mere reactions to the right, the proaction of complex assemblage sheds certainty for the sake of a more enlivening force field. As Crockett puts it, "being is energy transformation." From this dynamic materiality he infers— startlingly, in a time of much religious ennervation—that "theology is energy."[113] Energy matters, as we face not just our own burnout, but the exhaustion of multiple religious, democratic, and earth systems. What if the energy of transformation materializes in the nodes of entangled difference, in the contractions of sensuous intersectionalities? Instead of converting us to secular*ism*, the work of political theology lets us secularize boldly— even as we keep secularreligious faith with our deep energy sources.

So then, time being short, one more hint from Paul: The messianic contraction of time in 1 Corinthians 7 is replicated in 1 Ephesians 1:10 as the recapitulation of all things *en christou*. Here that messiah materializes the One, known or unknown, in whom "we all live and breathe and have our being" (Acts 17:28). All, without exception. We are folded into one another, members one of another, as all things become enfolded in a boundless embodiment. In the monstrous abbreviation of the now-time, the queerly contracted densities of the world can be lived without anxiety. In the recapitulative kairos of our moment, we may assemble ourselves anew right in the face of crisis. Apocalyptic geopolitics notwithstanding.

"One thing was clear to me: this time in the life of the country is a kairos time." Contemplating Trayvon Martin's murder, Brown Douglas concludes that kairos "is a decisive moment in history that potentially has far-reaching impact. It is often a chaotic period, a time of crisis." And then she adds, magnificently and as theologian: that now-time "is a time pregnant with infinite possibilities for new life."[114]

The kairos is still, and again, and ever, now. Might assemblage at the edge of chaos, deciding across critical difference, actually enact political inception? Or, in the stupor of optimism and the closure of pessimism, do we rule it out? And with it all the black-draped agonisms contracting present crisis with untold possibility?

"We owe each other everything." It all matters. What on earth is the matter with us?

2

EARTH

Climate of Closure, Matter of Disclosure

People are grabbing at the chance to see
the earth before the end of the world,
the world's death piece by piece each longer than we.
—Ed Roberson, "To See the Earth Before the End of the World"

Like those grabbing at the last chance, the poem's apocalyptic opening satirically grabs at our attention. But Ed Roberson's distinction of *earth*, which is not said to be ending, from *world*, which is, belies any simple closure. "The world" signifies a collective schema: human self-organization inextricably entangled in the nonhuman. So "the earth" evokes the planet, the earth that presents—is there "to see"—in its critical difference. And in the tensive time of the poem, earth presents the shortness of time, the fragility of world. The spatiotemporality conjured in that first verse turns tauntingly complex: "the world's death piece by piece each longer than we." But longer in just what frame of time-space?

> Some endings of the world overlap our lived
> time, skidding for generations
> to the crash scene of species extinction ...[1]

Roberson then juxtaposes the temporality of extinction to the scene of a crashing train, nation, land. The syntax of the crash scene conveys relentless speed, even as it hurls at us a landscape that caps mountainous stability with frozen stillness:

> That very subtlety of time between
> large and small
> Media note *people chasing glaciers*
> in retreat up their valleys and *the speed* . . .
> watched ice was speed made invisible,
> now—it's days, and a few feet further away . . .[2]

The "subtlety of time"—*subtilis*, "delicate," "finely woven"—binds us in the warp and woof of its fabric. Its interweave of diverse and shifting temporalities gives us no escape from the stunning speed of an actual ending. Not long ago I joined family at Glacier National Park only to realize that the very name will soon be a lament, an epitaph. I hadn't paid attention to U.S. glaciers before. Now I learned that of the 150 glaciers that existed in the park in the late nineteenth century only 25 remained. Global warming is taking them out. "The trend is consistent, there's been no reversal," comments Daniel Fagre, research ecologist and director of the Climate Change in Mountain Ecosystems Project, who has been the U.S. Geological Survey lead investigator for the Benchmark Glacier Program since 1991.[3] All of the glaciers will have died by 2030. "Outliving these glaciers is a little like outliving our children," Fagre says. "You pray it never happens."[4]

"Glacial pace" doesn't mean what it used to.

Chapter 1 considered the political notion of the exception, the emergency. Does glacier melt count as one? Outside of the park, and in the heat and speed of recent politics, the crisis of old ice will leave most cold. Not in Southeast Asia, where Himalayan glaciers provide not only the water for the sacred Ganges but for over a billion people. Global warming is affecting the Montana glaciers more quickly than others; they are "a bellwether" of planetary melting trends, including, of course, sea ice and ice sheets, the Arctic and the Antarctic.[5] (As I write, a *New York Times* headline flashes, "Alaskan Permafrost . . . No Longer Permanent."[6] It should all be melted by 2050, with its rising waters intensifying the methane-spiked feedback loop of global warming.) No exception then, this emergency.

Shortness of time moves at vastly different earth speeds. And that difference—so finely woven, a matter of so few degrees of warming—becomes critical:

> a subtle collapse of time between large
> and our small human extinction.[7]

Small in geotemporality, our extinction. No end of the earth. But no small matter for our *world,* is this speedy-slow contraction.

CRASH SCENE OF THE ANTHROPOCENE

If we *anthropoi* collect around critical difference—at different scales of geopolitics—the earth poses its difference from us precariously. Our politics may continue to disregard it, but earth makes up the polis, immanently, as the matter of us, even as it so geologically transcends politics. How then would the poem's "subtle collapse" be related to that contraction (sunestalemnos), that shortness of "the time that remains," we witnessed twisting into the kairos of a subtle ingathering? Confronting different fears, and in a very different time, Paul announced in this reading no end of the earth-*oikos,* but of "the *schema* of this world."[8] If, in this chapter, we zoom in on the earth as subject of political theology, it is not to announce that our planet is the one *doing* this theology. The planet is not a theologian, not even a closet ecofeminist theologian. Earth signifies the subject *matter*, the materializing life, of any politics and of any theology. Our subjectivities are in and of earth. Yet matter for a long stretch appeared subject to the anthropic exceptionalism: mind over matter, Man *über alles*.

In this time, matter is refusing to remain discretely enclosed in our world schema. It is icily mirroring us to ourselves in our species meltdown. And now would we learn to *mind* our matter? To question the terms of human sovereignty over the animal, the plant, the element—not to mention over fellow humans deemed animalistic, vegetative, or elementary?

This period of collapsing time frames takes different names. It is being widely now called the Anthropocene, but also, as we will see, the Capitalocene, the Chthulucene, the Ecozoic, perhaps the Ecocene. These time-names will help this chapter to mind the actual *earth* of this political theology of the earth. The planet does not supersede (Gaia forbid!) the multiple interhuman crises—of race, migration, sexuality, economics—overheating at their own speeds of crashing immediacy. Instead, it locates them all in the same home, the planetary oikos of our ecosociality—no takeouts, no exceptions. We cannot hold them in relationship all to each other unless we take in their common matter, indeed their material undercommons. To teach such earthen entanglement will not prevent great emergencies of climate change; but it fosters, it grounds, the emergence of a complex planetary public.

If our reading of political theology had led us to consider its notion of sovereignty as the power of decision in the state of emergency—the exceptional situation justifying exceptional authority—it is because in Carl Schmitt's politics of friend versus foe we see illumined certain currently acute motives. We do not in response deny either emergency or enmity. Instead, we stoke the chance that the unleashed U.S. enmity against the earth and its vulnerable populations, human and other, will provoke an answering emergence, self-organizing in multiplicitous solidarity. Antagonism can spark but not sustain it. Mindful anger, the outward lift of lament, burns with greater animation. In its amorous agonism does it not draw upon the *eros* and the animacy of all that surrounds us? Of all that mirrors us not only icily but meltingly, bracingly and embracingly, inhumanly or humanly? It contracts it all in the mud, *adamah*, of our mattering now-time. Such dirty love offers itself even to the foes of the earth. But it disgusts them.

In the present theological contraction, the earth will step forth neither as our problem nor as our solution. The earth is what is the matter with us.

WHITEOUT

In a multitemporal series of intersecting exceptionalisms, chapter 1 noted how an Anglo-Saxon exceptionalism morphed into the white supremacism that congenitally infects U.S. democracy. We glimpsed Benjamin Franklin's ecoracist dream: that as we brighten our land by clearing its dark forests we would also stop the darkening of our human population. If this White Earth fantasy was early and often interrupted by the dynamisms of demographics, labor demands, immigration, civil agonism, national parks, it continues to spook the future. Waves of reaction against the immigrant-racial other here and in Europe fed fascist potencies in the last century, returning to spook this century. But while German fascism manipulated identification with the local environment (*Blut und Boden*), the current white supremacism marches in lockstep with environmental denialism. Its triumph was thus permanently symbolized by an erasure: "Within moments of the inauguration of President Trump, the official White House website ... deleted nearly all mention of climate change."[9]

The sovereign assault on every environmental agency and law was pursued threafter with daily consistency and capitalist glee. White folk-house-earth. This happened just as the science had become ice-clear: To avert ecological collapse, we have a decade or so to change course (pretty much "the time that remains," as I write, to our glaciers, whether we change or not). By then we may face what in its beginnings is being written off to Muslim violence or immigrant criminality: the coming waves of millions, then hundreds of millions, of climate migrants, few of them white, fleeing the rising waves of the sea and the lengthening terms of drought. Christian Parenti has long warned of an answering "climate fascism."[10]

In the time of what Roberson calls "the world's death piece by piece," with its slow-fast temporalities, its grabs and its collapses, we cannot cease to ask each other: how do we in this moment recollect, rematerialize ourselves? How do we keep contraction from caving into closure? How to talk about these endings without triggering the cycle of apocalyptic catastrophism and despair, "Anglo-normative pessimism," and surrender? How might we tease *apokalypsis* into dis/closure? In other words, how shall we mind "the subtlety of time" and therefore of language? Consider how English, conventionally separating "Man" from the less than human, joins anthropic and Anglo-Saxon exceptionalism.[11] It doesn't have to.

Roberson has commented on the Black inflection of his own poetry: "I'm not creating a new language. I'm just trying to un-White-Out the one we've got."[12] As the crises of race and of climate become inseparable, they threaten a blanking out, a blanching (*blanche!*)—of our world. Whiting out: the race-face of the exceptionalist takeout.[13] Shall we try to unwhiteout the one we've got? While, in unspeakable irony, the white glacier caps melt forever?

CREATED IN THE EXCEPTION

For a political theology of the earth, the sanctifying sovereignties of language no longer find cover in a Word come down from heaven. Unwhitingout the language and legacy of the two Testaments (their poetics composed over a multiplicity of times by a multiplicitous writers' collective, no white folk among them), has been a long-time work of liberative theologies, North and South. As they mobilize messianic resistance against a Christianity aligned with imperial and neoimperial power, its "word of God" wobbles uncertainly. Its exceptionalism—of Christ and The Man—begins to pale visibly. In the "subtlety of time between," the logos seems to be finding itself redistributed. It has been showing up queerly intercarnate, in skins dark, female, animal, or vegetal, in saltwater and glacier, in earth and star, in every body of genesis.

And yet the biblical proof text for anthropic exceptionalism has always been the "dominion" symbol of Genesis 1:26. Human beings, and only human beings, as created "in the image of God," were thereby granted God-like dominion over all fellow creatures. This, at least, is the version that in modernity got robustly secularized. How effectively capitalized, politicized, and sanctified has been this word![14] And there is no denying that a certain sovereignty, in the sense of governance, does already inhere in the biblical pronouncement.

Nor can it be denied that this gifted *imago* of Genesis signifies, far from any supremacism, a creativity at the edge of chaos. For the purposes of a political theology of the earth, it is well to recognize that biblical scholarship reads the text in its political context. That is, the Hebrew narrators write from the experience of exilic subjection to imperial sovereignty. So, Genesis 1 can be said to reflect, indeed as postcolonial theory would put it, to "mirror and mock," the Babylonian religion of their captors as enshrined in the *Enuma Elish*, where humans, with no mention of fellow animals, are created to serve as slaves to the gods. Amidst the celebratory Genesis multiplicity, "dominion," as encoded by a colonized people, means that, far from being slaves, we humans—all of us, no exceptions—mirror divinity upon the earth. The explicit "male and female" displays another anti-Babylonianism, echoing the dedemonization of the feminized ocean goddess Tiamat.[15]

The text, composed when the nonhuman was still more dangerous to the human than vice versa, redistributes symbolic royalty: from its monopolization by the gods and their human representative to all humans. But is the point that we shall then do just anything with this gift of dominion? A warrant for twoleggeds to extract, exploit, and exterminate all others at will? To whiteout entire species every day? How bizarre thus to interpret the calling of the ones created in the image of that *Elohim*, the creator who on every symbolic day, upon witnessing the emergence, large and small, of each collective becoming of element or vegetable or animal—no exceptions—bursts into the "it is good!" refrain. Doubly odd, since no motive for this creative process is named other than that common, *cosmically* common, good.

Still, you might press, even when read as responsible stewardship, the *imago dei* can only be interpreted as a revelation of our *exceptional* human status. And I do admit that the human relation to the divine is here marked as qualitatively different, special, supercharged with affirmation, in that moment—who knows what size of time—before exile from Eden. The priestly narrator marks only our likeness to the creator. And yet the earth and the waters participate as invited in the creative process ("Let the earth bring forth," etc.), with outcomes praised. Earth and ocean seem to mirror more directly than the human the character of the creator—to create. (Humans are not here invited to "bring forth.") The creative source seems here to have endowed humans with a risky mandate: to deploy our gifts across the face of the planet, which can only be read, as the climate encyclical insists, in conjunction with Genesis 2:15: "Till the earth and keep it."[16] Modernity heard: "Exploit the earth and own it."

Just how exceptional a creature does the imago dei make the anthropos? An answer comes as the climactic content of that sixth day. Notice what is hidden in plain view in this hypertext. Following immediately upon the verse of dominion, the text offers, in detail disproportionate to this compressed chapter, our dietary plan. Humans get—*read* it—"every plant yielding seed that is on the surface of all the earth, and every tree which has fruit yielding seed; it shall be food for you; and to every beast of the earth and to every bird of the sky and to every thing that moves on the earth which has life, I have given every green plant for food" (Gen. 1:29–30). In other words, our honorific dominion comes down to this: we get to be vegans like all the other animals! And only so was it "all very good."

In the word according to Genesis, the human inception poses no exception to animality. We are emphatically not taken out of it; our impressive difference lives *within* animality. Our distinctive theomorphism only matters within the context of our creaturely interdependence beastial and vegetal. The subsequent narratives of sin (and climate change adjusted diets, as per the covenant with Noah) struggle with mounting human ambition and antagonism. But they do not erase the image and call of the cosmic common good: "the covenant with you, your descendants and every living creature" (Gen. 9:9f).

Despite manifold limitations, inconsistencies, acts of patriarchy and violence, the biblical narratives inaugurate no zero-sum game of Man versus Earth. Nor is matter ever reduced to a dull lump of rule-bound stuff or a grab bag of resources from which I except myself and mine. In this ancient-new materialism, the genesis collective is composed of mattering bodies—animals, animate forces, ecosocial animacies.[17]

POLITICAL ANIMALS

As to those animals, they have in theory been coming into their own. At least in theory's more transdisciplinary vitality—signaled by new materialism, affect theory, animal studies, critical life studies—theory in this century struggles promisingly toward the mattering of things, resistant at once to mechanistic reductionism and to anthropic exceptionalism.[18] Critical animal theory, for instance, has come to pass only through a certain autodeconstruction of anthropocentrism, indeed of the anthropic exceptionalism of deconstruction itself. But it did come. Derrida's late meditation on the *cogito ergo sum* (and his little cat) in *The Animal That Therefore I Am* (2006) undoes any *separation* of animal from human. And yet the *différance* between nonhuman and human animal is not thereby minimized. Au contraire, his alternative to anthropic exceptionalism is signified by "limitrophy," the growth of the limit: "Not just because it will concern what sprouts or grows at the limit, around the limit, by maintaining the limit, but also what feeds the limit, generates it, raises, and complicates it."[19] Indeed it "folds" and multiplies it. Difference is not washed out but made dense, complex. Contracted. No sharp line obtains between you and, say, Derrida's pussy cat muse, of course, because there is no way to extract "you" from your own animal body—breath, brain, genes, microbiome, vagus nerve, gut feelings, affect, and possibly cuddly softness or arch disdain.

In another parlance, we would recall that difference is itself never anything but a relation, even if it is a relation of mutual hostility, indifference, or ignorance. But this theorizing afresh of the human relation to the nonhuman animal cannot quite take place, in the west at least, without a bounce back through Genesis, and so to the *Other* side of the nonhuman: God. Christian theology came postbiblically to erect a transcendent separation from Him (*sic*), which simultaneously formalized our transcendent separation, in His image, from the animal. Derrida coined the term *divinanimality* to trouble that ultimate human barrier against (our) animality.[20]

Of course, the politics, even before the ecology, of our history of anthropic exceptionalism surges with the disavowed energies of an animality that returns as the repressed—and as the repressive. Derrida's late lectures on *The Beast and the Sovereign* (2001–2002) question the political in part by way of a critique of Carl Schmitt's omnipotently charged authoritarianism, as simultaneous appropriation and denial of its own beastliness. Such "zooanthropological" beastliness operates outside the law, whether above it in the aggression of the sovereign, or below it as projected onto his victims.[21] While Derrida never develops an explicit political theology, as Clayton Crockett demonstrates, his "messianism without a messiah," his "democracy to come," performs the prophetic theological repetition of the theology we began in chapter 1 to solicit for political theology as such. Crockett's conjunction of deconstruction with the new materialism recycles indispensably the ecosocial, and only so *political*, potentiality of divinanimality.

For a political theology of the earth, animality registers first of all as the counterexceptionalist embodiment of our interface with the nonhuman. Our kinship with the entire bestiary of the earth—that is, our common animality—thus tests the commonality of any good we might materialize, retrieve, protect, enjoy, or lament. For these affects that keep what matters mattering to us come seeded deep in our beastliness. In *What Animals Teach Us About Politics* (2014), Brian Massumi lifts from zoology "the lived importance of animal-human relations." In an arena "thick with corporeality and affective complexity," this Deleuzian affect theorist is lead to "an active ecology of a diversity of animal practices, in a creative tension of differential mutual inclusion."[22] Creative tension of differential mutual inclusion: yes, that captures the nonunitary contraction we seek. He offers an indispensable negative thesis: "Do not presume that you have access to a criterion for categorically separating the human from the animal." Often language is used to mark that wall (we name *them* after all), and yet "animal play, in fact, produces the real conditions of emergence of language." I couldn't help thinking about this animal play during the women's marches of 2017: women and friends swarming, crowded close, wearing knitted pussy caps, mingling laughter with rage, holding up humorous signs: "Ninety, Nasty and Not Giving Up"; "Electile Dysfunction"; "So Bad Even Introverts Are Here"; and, of course, "No Planet B." Another example of the political play of us animals: our widespread dependence for an *alternative* "alternative truth" on the comic genius of Stephen Colbert or Trevor Noah.

In his analysis of religion and affect theory, Donovan Schaefer has tracked the evolutionary biology of our affects—which he makes clear signifies their animality—as the explanatory framework for religion and its relation to power.[23] The neglect of affect and the preoccupation with language has left us perilously unprepared, he shows, for explosions of such religiopolitical forces as U.S. racism or international Islamophobia. And, one must now add, for their successful electoral amplification through pale male antagonism. If those white affects seem in motivation alien to even the most aggressive nonhumans, their glandular intensity does not. And their spread across classes, even genders, testifies to the danger of neglecting, whiting out, the animal affectivity at the base of the human. We do not need Schmitt to tell us how inept liberalism remains at cultivating identity-forming *democratic* passions. But we might need Aristotle to remind us that we are "political animals."

TAKING THEM OUT

If our difference represents no exception to animality, we cannot *take ourselves out* of our creaturehood. That etymological signal of the exception, its "takeout," now startlingly flashes from the climax of Roberson's poem. Here he meditates on the arc of history that links our time to the paleolithic:

> ... All that once chased us and we
> chased to a balance chasing back, tooth for spear,
> knife for claw,
> locks us in this grip
> we just now see
> our own lives taken by
> taking them out. Hunting the bear,
> we hunt the glacier with the changes come
> of that choice.[24]

"We just now see . . ." *Excipere:* taking out those objects of our once balanced predation has now lead to what Elizabeth Kolbert calls the sixth great extinction spasm. These words of ecologist Paul Ehrlich are emblazoned in stone capitals at Stanford University: "IN PUSHING OTHER SPECIES TO EXTINCTION, HUMANITY IS BUSY SAWING OFF THE LIMB ON WHICH IT PERCHES."[25] Human time so long, one-hundred-ninety-five-thousand-years-long, is contracting now with whiplash speed: "Our own lives taken by taking them out." In this time of collapsing times, can collective sight shift, in some critical mass of eyeballs, from voyeuristic grabbing to dawning awareness? "Of the changes come of that choice"?

So then the anthropic exceptionalism has, with a certain sovereign consistency, brought us into this new time increasingly called the *Anthropocene.*[26] It names the era that has succeeded the twelve-thousand-year climate stability of the hospitable Holocene, the only known context for human self-organization as what we call civilization. A growing proportion of geologists are using the term as a formal designation.[27] At the same time, many stress that the earth is always changing, and humans have been affecting the planet practically since the beginning of the Holocene (so the Anthropocene cannot name a mere exception).

Other critics fear that the term *Anthropocene* fosters a new anthropocentrism, even faith in a techno-redemptive answer to climate change.[28]

Against such critique, Tim Morton declares that "Anthropocene is the first fully anti-anthropocentric concept."[29] For him, "The end of the world is correlated with the Anthropocene, its global warming and subsequent drastic climate change, whose precise scope remains uncertain while its reality is verified beyond question."[30] By this, Morton means that "the end of the world has already occurred." With that rhetorical punch, Morton wants "to awaken us from the dream that the world is about to end, because action on Earth (the real Earth) depends upon it."[31] World thus designates a particular schema. But is it so neatly behind us? Shall we then reverse the poem and see the earth *after* the end of the world?

Other ecological theorists propose the term *Capitalocene*. It captures unambiguously the obscenity of the current world schema. From this point of view, global capitalism in its neoliberal planetary format is indeed bringing on the end of our world. As Parenti analyzes the Capitalocene, it does not take aim at a simple economic globalism, as having cut loose of politics. Rather, and crucially for a political theology of the earth, "it is the state that delivers nonhuman nature's use values to capital."[32] With subsidies, tax breaks, and police backup, modern politics in its sovereign territoriality serves up nonhuman nature to capital accumulation. And so we face a planetary emergency that "requires immediate action on a truly massive scale."

In other words, we must divest from the sovereignty of the present politico-economic world-scheme. Capitalism, with ever less pretense to trickle down to any common good, serves its exceptional 1 percent; and in its extractions, it abstracts itself, *our*selves, from the earth itself. The melting of glaciers, desertification of earth, acidification of ocean, erasing of shorelines, mass migrations—all these count in economics as mere "externalities."[33] *Taken out* of the calculation. As does the well-being that fails to reduce measurably to profit, of earthlings, human and otherwise.

Still, Morton has a point when he argues that the term *Capitalocene* misses the mark. If capitalism were the sole problem, Soviet and Chinese carbon emissions would have added nothing to global warming. But, as I write, there is encouraging news from China, which may, despite its double legacy of communist and capitalist ecological indifference, be beginning to fill the widening vacuum of U.S. leadership. Given the aggressive capitalism of China, this would not disconfirm Morton's point. Moreover, reading capital as the unitary planetary evil lets various precapitalist sovereignties of imperial violence, including the Christian ones, off the hook of history. If Connolly's "capitalist-evangelical resonance machine" remains in play, the simplification of the world scheme to economics as the single cause will again miss the intersectional force of affects, motivations, and values powering neoliberal capitalism. In the interest of the political as the self-organization of complex ecosocial systems, we can elude any reductive explications of the problem—even reduction to the capitalist machine of reduction itself—and we can do so without underestimating the planet-degrading totalism of late capitalism. So why not just avow the Capitalocene as primary, not sole, driver of the Anthropocene?

It is not often enough realized how the speed of the Anthropocene extinctions mirrors, across alien scales of time, the temporality of capital itself. It is the high-speed plasticity of the current form of financialized capitalism that theologian Kathryn Tanner's recent Gifford Lectures unveil.[34] With this rapid-response plasticity, financial institutions have the capacity to absorb a history of multiple causes into their management of the future. It projects the terms of short-term profit into a homogenized future.

Indeed, the high speed of capital contrasts triumphantly with the slower temporalities of the planet it takes as its stable space and its predictable time—as its real estate. But in the real state of the earth, the "subtle collapse" of times, anthropogenic and capitalogenic, is becoming visible. And it shows that, as the global economy has already driven the earth's ecology out of the Holocene, the temporal schematism of steady progress in time becomes steadily more delusional. As they ever do, multiple destabilizations, human and inhuman, take place. But now they work increasingly together, in a new ecosociality of crisis, within the shaking scene of the Capitolocene. So, for instance, the international uptick in populisms is not earth-friendly; but, as Michael Northcott argues, it poses new contradictions within the smooth unfolding of a sovereign capitalism.[35] The authoritarianism gets fed by the economically driven alienations of European and U.S. populations from their originally place-bound and now readily racialized, identities.

Tanner does not herself focus upon the ecosocial or political implications of the global economy at this moment. Rather, she has significantly advanced the economic analysis whereby responsible Christian thought can sharpen its critique of capitalism. Without a grasp of this high-speed adaptability of global economics, political theology will lose its grip on the convulsive present. Tanner ultimately offers a vision of "radical time discontinuity, promoting expectations of radical disruptive change."[36] Building on her prior exposition of "economies of grace," she unfolds a theology of arduous engagement with the reigning schematism. Given the person-shaping power of the model, she counters it with the capacity of Christian beliefs and practices to help people resist the dictates of capitalism in its present, finance-dominated configuration. The high-speed plasticity of financialization fosters a temporality with no investment in the longer-term future in which species, ours included, unfold. Its short-term gains by their very flexibility trap its subjects in a continuous present and therefore in themselves. In other words the economy performs the Capitalocene opposite of the contractive relationality of the Pauline kairos. So then the foreshortening of time may mean mere surrender and sellout, or it may mean awakening to the coordination of financialized time with the end of the world. And therefore to an alternative.

Dauntless activist and author Naomi Klein names what is at stake: "Our economic system and our planetary system are now at war. Or more accurately, our economy is at war with many forms of life on earth, including human life."[37] Advocating neither for Anthropocene nor Capitalocene, Klein continues: "What the climate needs to avoid collapse is a *contraction in humanity's use of resources;* what our economic model demands to avoid collapse is unfettered expansion. Only one of these sets of rules can be changed, and it's not the laws of nature."[38] No exceptions there!

When Klein calls the alternative to capitalist expansion and collapse "contraction," Paul's sunestalmenos is surely the furthest thing from her mind. Nor, for all of Paul's asceticism, did he have any contraction of global resource use in mind. But as the present project draws upon the disruptive political theology of a "weak messianism," we might let the kairos of an ecosocial inception echo between Klein's rousing lines: "Fortunately it is eminently possible to transform our economy so that it is less resource-intensive, and to do it in ways that are equitable, with the most vulnerable protected and the most responsible bearing the bulk of the burden."[39]

Does this eminent possibility ring noncredibly hopeful for this now, just a bit later? Certainly—if one confuses possibility with probability. I hear in Klein neither optimism nor pessimism but an effectual secularization of the Hebrew prophetic hope. It came always edged with ecopolitical protest: "therefore the land mourns, and all who live in it languish; together with the wild animals and the birds of the air, even the fish of the sea are perishing" (Hosea 4:3). It is only from the perspective of what is possible that judgment rings out in protest. Judgment demands justice because (in a slogan prematurely aged) another world is possible.

In her earlier *Shock Doctrine* (2007), Klein had investigated the global manipulation of emergencies by unregulated, indeed sovereign, capitalism. Her analysis exposes in economics an analogue to Schmitt's theory of the emergency as key to *political* power.[40] In *This Changes Everything*, Klein warns of a "frenzy of new resource grabs and repression."[41] If the managerial delusion of a continuous present no longer quite holds, even with the managers, intensified exploitation and extraction will exponentially increase the whiteouts, the extinctions, as by-products. To these shocks of neoliberal capitalism, she juxtaposes this revolutionary possibility: that "climate change can be a People's Shock, a blow from below. It can disperse power into the hands of the many rather than consolidating it in the hands of the few, and radically expand the commons, rather than auctioning it off."[42]

Such an outbursting of the commons seems to suggest—in its "radical time discontinuity"—a dispersion effected, like a supernova, like grace, by the ingathering intensity of a public contraction.[43]

MORTAL CRITTERS ENTWINED

Is this alternative ever more upon us, this chance of a choice, or choice of a chance, to call forth a fresh meaning of "we"? To contract together in a new earthling *intersubjectivity*? Of course, the old human may close down rather than contract. Or instead, in resistance to the delusional infinity of capital growth, we may press open a commons of dis/closure. We might avail ourselves of that residual plural form of the singular noun *commons*, of *common good/s*, enabling coalitions tested by the critical difference of their planetary undercommons.[44] In particular ingatherings, complicating our localities at their limitrophic edges, that plurisingular commons (in the evolutionarily refracted, profanely mattering image of the plurisingular Elohim) forges, indeed it animates, the chance of an unprecedented earth public.

Our animal kin cannot do our politics for us. But we cannot now do the political without them—which is to say, without admitting their humming, roaring, barking, buzzing input. Is a chaos of the nonhuman, animal, vegetable, mineral, irrupting just where the anthropic exception maintains its lower boundary? If the genesis of a public capable of a viable earth future takes place at the limits of the human, it assembles only at the edge of chaos: in a *creatio ex profundis* abysmal with loss but fluent with possibility. This is not a story of origin but of inception. And when would that take place—if it is to happen as our planetary place at all—except now?

If the present time may be, in mindful critique, dubbed Anthropocene, it is then a present named for its devastating epochal novelty. But it is widely associated with a technological inception (like the movie after all). It may then cancel its own possibility, the *kainos*, which is to say, "the new"—the root of the suffix *cene* attached to both of these epochal designations. So Donna Haraway proposes the *Chthulucene,* combining the *chthonic* with the new: "This Chthulucene is neither sacred nor secular; this earthly worlding is thoroughly terran, muddled, and mortal—and at stake now."[45] She is invoking corals, octupi, spiders, as well as her old canine companion, in a great swarm of "kindred critters." "To renew the biodiverse powers of terra is the sympoietic work and play of the Chthulucene."[46] The element of animal playfulness animates all her writing. She would not want these powers sacralized under the banner of any theology. No more would we, if the sacred signifies some transcendence, as exceptionalist as its techno-redemptive secularization., of the mattering earth,

Haraway's *sympoiesis*, a becoming-with or making-together, is evolved in proximity to Lynn Margulis' evolutionary theory of *symbiogenesis* and is meant as an alternative to *autopoiesis*, the standard designation of self-organizing systems.[47] Sympoiesis helps our definition of the political in terms of self-organizing collectives, which should not be misunderstood in terms of self (*autos*)-contained systems. The self-organization that might foster the animate, agonistic earth commons would express a "self" not of autonomy but of entangled difference. The human schema of such sympoiesis would then magnify rather than whiteout those "biodiverse powers of terra." "Specifically, unlike either the Anthropocene or the Capitalocene, the Chthulucene is made up of ongoing multispecies stories and practices of becoming-with in times that remain at stake, in precarious times, in which the world is not finished and the sky has not fallen—yet."[48] (Chicken Little, after Aesop's wolf, may still have a chance.)

We have not escaped the collapse of times, but we are sniffing out time's own subtlety. And in our multispecies contractions we find no escape from each other: "We are at stake in each other."[49] We become-with—or not at all. In the indeterminacy of our human and nonhuman entanglements, we cannot determine with certainty where the limit lies, the limitrophic margin between what climate change has already determined and what we may still regather, regroup, reassemble in the earth. Anthropocene and Capitalocene narratives make perhaps too much of our anthropic agency—and so, ironically, too little. For it is human hubris that is leading our abdication of responsibility, the ability of our shared agency amidst precarity.

Working at that time limit whose subtlety neither optimism nor pessimism can read, Haraway offers the great and humble strategy of "staying with the trouble." Confronted with what cannot be denied, fixed, or affirmed, "staying with" names a psychosocially wise maneuver for any crisis of difference. It obliquely resembles what in a Christian register Shelly Rambo has called "remaining."[50] In the light of trauma theory, she observes the lack of any final fix, any assured future, but a way in the aftermath of loss to "improvise along the tangled lines of what it means to remain."[51] She refuses a triumphalist leap from crucifixion to resurrection; and so she rescripts the very scene of the resurrected messiah in terms of "the bodies of those living," where "the surfacing and crossing of wounds presents a vision of interconnectedness in perilous form."[52] We gain in this theological griefwork another lens for reading "the time that remains." Indeed we may glean from such remaining-with, like Haraway's staying-with, a welcome transcription of the last chapter's intersectionally amorous agonism in its resistance to the sovereign antagonisms of triumph and vengeance.

"Staying with the trouble requires learning to be truly present, not at a vanishing pivot between awful or edenic pasts and apocalyptic or salvific futures, but as mortal critters entwined in myriad unfinished configurations of places, times, matters, meanings."[53] Haraway deploys this practice in exacting relation to her Chthulucene narrative: "The chief actors are not restricted to the too-big players in the too-big stories of Capitalism and the Anthropos, both of which invite odd apocalyptic panics and even odder disengaged denunciations rather than attentive practices of thought, love, rage, and care."[54]

The activisms and interactivities of live, if troubled, critters return us thus to the matter of time's scales. But time readily gets snared in the human schematisms that frame too big a story of Us. Apocalypse read as closure, whether secularized as Manichean geopolitics or as progressive despair, does enclose all of history in its predetermined End.[55]

If, however, in staying with the troubles one begins to sense a hopeful energy, in solidarity perhaps with Katongole's "hope in ruins" freed from any certifiable object-future, one misreads Haraway. She keeps hope off the list of attentive practices. She does not distinguish it from salvific optimism. "Neither despair nor hope is tuned to the senses, to mindful matter . . . to mortal earthlings in thick co-presence."[56] As the future skids into darkening shadows, we are hearing more of such non-nihilist, life-affirming relinquishment of hope. No doubt that abandonment has often been the case with hope. (Hence T. S. Eliot: "wait without hope / For hope would be for the wrong thing.")[57] In theology it has taken the strong recent form of Miguel de la Torre's *Embracing Hopelessness* (2017). Hope, clinging to its object, may displace the present. An image, even in the most socially just of imaginaries, may replace the place-time of embodied, contextually lived life. Hope is susceptible to the "fallacy of misplaced concreteness" (Whitehead), confusing the present—embodied—potential with an abstraction.

One might nonetheless wonder: how is hope any guiltier of displacement and stereotype than love or rage, in which Haraway finds "germs of partial healing"?[58] Moreoever, what love or rage does not also implicate a future? Except in the practice of Buddhist nonattachment—in which case love and rage are therapeutically emptied out, along with hope—what attentive relation does not love and rage beyond the present, which is always past by the time you love it or rage against it?

Of course, we must all choose which of the contaminated words in our troubled secularreligious vocabularies we will stay with. Staying with the troubles, which certainly implicates future troubles, may require something very like "hope"—the embrace of possibility—if it means to stay-with for more than one moment. In other words, if time is not empty, continuous chronos, each present contracts at once its past with its potentiality. So if we then disallow hope, we may miss and so fail to embrace the possibility tucked in this very present. For in that now-time the present tense is a contraction of our past with our possibility—as encoded in Winter's "melancholic hope" or Solnit's "hope in the dark."

So, instead of confusing it with optimism, I would rather hold hope's feet to the fire of its most amorous—most *remaining*—desire. Its present tense then contracts our past with our possibility. And every possibility is a lurch into a future present. But when hope ex-cepts itself from the thick copresence of what matters, pronounce it "hype."

ECOCENE INCEPTION

Haraway's barely utterable *Chthulucene* brings home the unnameability of this time that begins as an ending, that ends the holism of the old-"new" kainos of the 11,700-year Holocene. According to Thomas Berry, the inspiration of a persistent tradition of multireligious ecoactivism, we have terminated not just the Holocenic time of climate stability but "the Cenozoic sixty-five million years of geo-biological development. Extinction is taking place throughout the life-systems on a scale unequaled since the terminal phase of the Mesozoic Era."[59]

Oh those scales: "the world's death piece by piece each longer than we." Yet Father Berry named the new era the "Ecozoic," as though the catastrophes we are perpetrating might in being recognized catalyze a mass mindfulness. The activist scholars who lead in his wake, especially Mary Evelyn Tucker, Heather Eaton, and John Grim, elaborate on the basis of a multifaith ecology a "journey of the universe."[60] It contracts the narrative of earth's multitemporality to the possibility of radical human transformation. Its cosmological story can be read as naively hopeful, especially if one does not read it. Entering the Capitalocene/Anthropocene debate, the sociologist of biodiversity Eileen Crist asks why we should not then "choose a name whose higher calling we must rise to meet?"[61] She opts for the Ecozoic.

The call is as steep as the fall is deep: the event of the collapse—the People's Shock—could (might not, but could) inaugurate the Ecozoic as "a new mode of being on the planet" that "can be brought into being only by the integral life community itself."[62] Speaking of higher callings, the papal *Laudato Si'* also calls for an "integral ecology." Integration signifies here no totalizing oneness or unitary expectation but rather an appeal to complex human self-organization in mindful interdependence with the biosphere. To integrate in this sense is to contract in diversity.

And if the "higher calling" is not to except itself out of the thick copresences of its Chthulucene kin, the hope of the Ecozoic cannot be for a one-shock blossoming of emergency into emergence. Emergence can only happen spasmodically, uncertainly, in an experimental agonism that remains-with a becoming multiplicity of practices. But in its dense localities its emergent public can no longer lose touch with its planetarity—with the epochal width, depth, and height of its contracting space-time.

To better designate this possibility amidst the scene of the debate, we might also speak upon occasion of the *Ecocene*. For politically theological purposes, let it surface the Greek original of *eco*. Let it claim the space-time of the common *oikos*, "household" meant *eco*nomically, as in Francis' "common home," and to conjure the precious, the precarious hospitality of the earth. For a globalism of private properties, the earth as "home" only exists as broken and leveraged into real estate. If earth's hospitality for us is in jeopardy, the kairotic contraction also reveals its wildness and fragility. This disturbing, ever less familiar *earthome* needs, unlike Mommy Earth, to be nourished if it is to nourish us. It will require sensuously mindful attention: as, for example, in the organic garden students at my theological school cultivate. They pass on some of the yield to the local food bank, enjoy experimenting with heritage, biblical, and regional crops, learn survival skills for staying-with unhoped-for-troubles; they host rituals of blessing, mourning, and *remaining*.[63] The Ecocene earthome at once warns and invites.

In an ancient example of public nourishment, a crowd in the wilderness received the impossibly multiplied loaves and fish. You can hear the story as cliché-miracle of natural law suspended by transcendent sovereignty—an interventionist exception. Or as told in a text whose privileged genre is the parable you might hear otherwise. The feeding of the hungry multitude reads as a parable of human/nonhuman integration under desert conditions of greatly reduced resource use and high-stakes sensory sympoiesis: an ecosocial inception.

In the becoming possible of what had been impossible, the new does happen. If it happens in the Ecocene, it takes place—and just *in time*—under anthropocapitalochthulocenic conditions.

CONTRACTED INFINITY

If the naming of this contracting-and-yet-opening-time partakes of impossibility, a political theology of the earth takes part always already, we had noted, in negative theology. Not only do we not know the sweep or time of what will happen, theology ceases to posit any certitude of God or providence. Thus freed of any positivist hope, this mystical edge of contemplative religious practice has quite long-standing experience in "staying-with" the unnameable, even, or especially, as we struggle to name it. That sometimes means remaining with unspeakable loss. Trauma theory in the late twentieth century demonstrated that suffering does not follow a time line of repair, as if it were the case "that in time," as Shelly Rambo puts it, "one can just get over or beyond it." (Remember Benjamin's traumatized angel of history.) "Both individually and collectively trauma marked a problem of living in the present, given that the past was still a 'living' and intrusive reality." The accompanying "breakdown of assumptions about the knowability of experience" was reflected "in the negative and apophatic language around trauma" as unspeakable, unknowable, unassimilable experience.[64]

In minding the emerging crisis of the earth, we may name the unassimilable, the irreparable, as ecotrauma. Yet we pause now and again to note the strain of naming. So *Cloud of the Impossible,* which lands amidst the unspeakable abuse of earth's hospitality, is named after a theologoumenon of the fifteenth-century Nicholas of Cusa. It stands in a line of interpreters of the Exodus figure of the "dark cloud" who evoke the apophatic encounter with the unknowable—sometimes liberating, sometimes ecstatic, sometimes traumatic. Here a mystical dark-out of the infinite may also un-White-Out finite possibility. Long before Copernicus, Cusa recognized the Earth as neither fixed nor at the center, as the entire universe of fixed and separate bodies comes to be decentered and thereby interdependent. "The blessed God has so created all things that when each thing strives to preserve its own being as a divine office, it does so in communion with other things." Here he cites a pre-Christian tradition that "calls the world an animal." A universanimal?[65]

This animate universe is created "as much as possible like God." In other words, not the anthropos but the animate multiplicity of the All unfolds in the imago dei: here God's face is recognized as "the natural face of all nature." So this God appears counterexceptionally not just to us and like us. Cusa exegetes the imago dei divinanimally: "if a lion were to attribute a face to you, it would judge it only as a lion's face; if an ox, as an ox's; if an eagle, as an eagle's."[66] Cusa's universanimal is read "in learned ignorance" as an infinity—not identical with but "as like as possible" to the divine. All, as the limitless materializing multiplicity of the world, is complicated—*complicans*—"in and as God," in a holographic manifold of enfoldings and unfoldings.[67] And intriguingly, in contrast to the purely negative (nonfinitude) of the divine infinity, the cosmos is distinguished as a "contracted infinity." All are enfolded *in all*—without exception—through the "contraction of the universe in each creature."[68] It is this *contraction* that made it possible for him to think the impossible astrophysical *expansion* of space. The Earth ceased to be the (one exceptional) center. For no one body would form a fixed center for this universe centered only in the omnicentric *complicatio*.

Cusa's decentering of the universe radically relativizes the human perspective. But in the following centuries' antagonism between religion and science, the relativized earth morphs, to the contrary, into a modern object full of objects, knowable and usable by Man. And modernity would soon enough trade the infinity of contracted communings for an endless human expansion.

Otherwise, who knows, perhaps—in a poetry that knows its own unknowing—the "subtle collapse of time between large and small" would mean something other than "world death." Something like the kaleidoscopic interplay of diversifying animacies. Something like the multitemporality of the macrocosm in the microcosm. Something more musical in the disparate scales, the incomparable tempi, of space-time. Collapse might then—and yet—in its contraction, its interthreaded *subtilis*, elude mere end.

LABOR OF A COSMIC ECOLOGY

Even now? As time goes into a stranger slow-fastness, will we, this species that learned to disdain both our own animality and the world's, deny and amplify our moment's ecotrauma? A capital-obscene done deal? Since the time is short, a collective kairos—let alone a coming epoch of mindfully ecocenic unfolding—will seem quite calculably impossible. Yet at the same clouded time, another contraction, ingathering, folds us mindfully into the agonism of the becoming earth.

If, in Roberson's language, "we just now see," then a more common good has a chance. In time? Is this then the time not of a knowable outcome, not of the telos we know how to hope for, but of the "untimeliness of the infinite surprise," even of "a messianic extremity" (Derrida)?[69] So the present question remains: if its ecopolitical schema gets here and there tuned *theologically*, might its untimely but deeply temporal possibility be enhanced? If theos now refigures itself in divinanimal configurations of earth, if it melts with the glaciers and rises with the seas, suffers with the traumatized and moves with the mobilized . . .

Where, for instance, political alliance for ecosocial justice wants or needs to reach into a variously Christian public, we might render the Pauline figure of the "household of God," *oikos theou*, quite literally *as ecogod*. Paul's use of the phrase interchangeably with the *ekklesia* precedes any sense of religious institution. Its habitable sociality performs the radical interdependence that was first materialized as the ingathered "body of messiah."[70]

Of course I anachronize. Paul himself, traveling the imperial roads between urban centers, rarely thought of the nonhuman world around him. As in 1987, and with reference to ecology, Jacob Taubes joked that Paul had "never seen a tree in his life. He traveled through the world like Kafka. I know types like that in Israel."[71] Paul's legacy has since undergone ecotheological supplementation.[72] Brigitte Kahl exegetes him in league with "Gaia and the Cosmic Ecology of New Creation." "Paul's empire-critical theology"—his political edge—"has a profoundly ecological dimension as well."[73] If Paul's currently contracting household is to deliver earth-minding energies to the planetary collective, it will be in communities learning how to stay with the troubles. Troubles of trees and glaciers included. The ecogod comes embodied, enfolding and enfolded, in the entire becoming creation. The troubled biblical hope of the "new creation" did not and does not expect a supernatural supersession of this world, a new creation out of the nothingness. It embraces the possibility of the radically mattering renewal of our earthome.

So then we take another Pauline hint. In his eschatological reflection on the now-time, he writes to those gathered in Rome, "We know that the whole creation has been groaning in labor pains until now. And not only the creation, but we ourselves, who have the first fruits of the Spirit, groan inwardly while we wait for adoption, the redemption of our bodies" (Rom. 8:22). "Until now," as the entire time of "the creation," is one long time to be in labor, aeons longer than biblical writers could have guessed. These birth agonies undo a certain Christian schema that, despite the labors of ecofeminist and all critical exegetes, persists: at the point of origin, the creation is "taken out" of mere nothingness, finished in "seven days"; at the end of the time line, the "end time" yields to the exceptionalist temporality of a top-down *new* creation from nothing. And with this imaginary grew the treatment of the creation as mere matter, meaningfully enlivened only for and by exceptional Man, and the sense of salvation as of souls taken back out of the body into timelessness. This temporality comes undone in the aeons of ongoing cosmic labor.

If we had time to read this agonism of creation properly, we might begin with the feminist critique of the ancient imaginary of the all-too-material *mater*, who is created, earth and all, and ranked beneath the immaterial Father, who creates. Really, we would not cease with our old feminist resistance (refreshed by recent assaults on the right to choose) to any natalism. And, with new ecological urgency, Haraway's "make kin not babies" blasts through.[74] As do queer worries about "reproductive futurism" (Edelman). Does the reproduction of the schema of civilization as straight line of heteronormative succession here find cosmic sanction? These normative effects will not disappear, however alien to Paul's point, with its mixed metaphor of birth and adoption, as well as to his discouragement of marriage and reproduction.[75]

And then we might interject Hannah Arendt's Jewish-Augustinian challenge of "mortality" with a non-natalist "natality," by which she resists the emphasis of philosophy and particularly Heidegger upon "the horizon of death."[76] Natality did not make Arendt optimistic (hence we find her warning, already in 1946, of the threat of a new "white supremacist" fascism driven by crises of immigration).[77] But each moment of beginning anew, of action, is touched—almost kairotically—by the wonder of birth. Her reading of Benjamin helped her to interrupt the continuum of time in a remembrance of the past that makes for a future.[78]

And back behind Benjamin, in a deep past textual and cosmic, we might sense the originative contraction of the kabbalistic *tzimtzum*, the contraction of the Eyn Sof, the Infinite, into itself so as to make space for the worlds, the creatures, the finitudes of creation.[79]

None of the aforementioned interventions would manage to translate the birth pangs of the world into either a reassuring or an expendable figure.

With Arendt's natal appreciation of Benjamin, we happen to find ourselves again within the rhizome of secular Jewish expressions of political theology. That conversation shapes, as we saw in chapter 1, Agamben's *Pauline* reading of Benjamin's *Jetztzeit* as the messianic now-time of a great contraction. So how can we not draw the analogy, however unintended, of that sunestalemnos of time with the *contractions of birth*?

The kairos from this perspective performs its ingathering as a contraction of the world-womb. The birth pangs do not signify any collapse of time into mere death. Yet this cosmic agonism does not assure a fruitful outcome. Its struggle can hardly be distinguished from the traumatisms of miscarriage. Stillbirths of Christianity, of hope and justice, of democracy and socialism, do not bode well for progress through pain to fruition. But neither do they rule out events of earth-birth through and beyond these agonies.

For a political theology of the earth amidst the pain of its present contractions, it matters that a redemption *of* our irreducibly earthbound bodies, not *from* them, was meant by Paul. As it was by earlier biblical prophets of the new materialization of heavens and earth. With Paul, it was anticipated in the context of an asceticism alien, on the one hand, to ancient repronormativity and, on the other, to otherworldly disembodiment.[80]

The parable of the creation groaning in labor seems to enfold our agonies in an embracing cosmic process. The human and the terrestrial contract together in the animal dynamism of an integral cosmic body. We could ask: Whose womb? That of the creation. But who is *she*?

Oddly the poet of Job has Elohim ask that question rhetorically, from the whirlwind conjured by Job's agonizing. "Who shut in the sea in doors, when it burst out from the womb? When I made the clouds its garment and thick darkness its swaddling band" (Job 38:8–11). The dark cloud of all negative theology here softly enfolds the neonate *mayim*, the fluidity of "the heavens" (*hashamayim*, "sky") and earth itself. Who captures this bursting force, weaves its garment and swaddles it? Mom? Midwife? The questions remain in biblical patriarchy unanswerable.

If, as John Wesley insists, that God, who "is in all things," is "in a true sense the Soul of the universe,"[81] it takes only a minor lick of anachronism to infer the following: unless the divine spirit exists in a Cartesian separation from its own matter, the womb can only be that of God. Later, some would articulate the metaphor of the universe as God's body.[82] Of course all biblical hints at a cosmic maternity, at all matter contracted in the divine womb, fall silent in evasion of the mother goddess—or, later, of pantheism.[83] I would rather be silent in avoidance of any God-capture. The poetic pangs of the ecogod and her immanence to the pains of all of us creatures remain hints. Hardly conceivable, linguistically or sexually.

In this multiplicity of becomings before and beyond all conception, there take place creaturely inceptions of mattering—of embodiments beneath and beyond our knowing. A political theology of the earth may on occasion avail itself of fresh hints of an ecodivine *intercarnation*.[84] But always it minds the contraction of the earth in happenings beyond one planet, in that unspeakably immense cosmos—not so cozy as "the creation" has signified. In the impatience of crisis we neglect that largely dark universe of mostly dark matter. Then we are at peril of losing the width and wonder not just of "the heavens and earth" but of the context of natural and therefore of climate science. The matter that swaddles us in our planetary flesh flows as the energy of obscurely distant timespaces and of primordial outbursts, big bang or big birth.

MATTER'S MESSIANIC MOMENT

If earth is the matter with us, we had best read its matter mindfully. And then we do not abstract it from its contractedly infinite universe, but wonder at the intimate manifold it has folded into itself. Each single contraction, any throb of critter, might invite our attention.

At least in theory. When, for instance, a theology of apophatic entanglement (impossibly clouded) offers itself to political theology, it tugs with it the apolitical materiality of a microcosmic entanglement. It carries in its cosmology a scientific exploration happening just where science verges on silence: of how quantum entanglement unveils the immense scale of an interconnectedness that contracts to the teensiness of the electron.[85]

Why go there now, in the midst of the harrowing political mesocosm of a particular planet? To flee earth's messy agonisms by circling between the great above and the great below? Always a temptation. But a terrestrial cosmology—inscribed in every indigenous mythology—may be what lets us tune to the *matter* with us. It may help us undo, from the microcosms of materialization up through the immensities, the exceptionalism that takes us out of our creaturely condition. Then the matter we warm-bloods share with glacial ice might collect us as a really big universanimal "we."

Hear a voice from the undercommons of matter itself—call it the *undercosmos*—speaking up for the most common element of all, the electron: "Even the smallest bits of matter, electrons—infinitesimal point particles with no dimensions, no structure—are haunted by, indeed, constituted by, the indeterminate wanderings of an infinity of possible configurings of spacetimemattering in their specificity. Entire worlds inside each point; each specifically configured.... Finitude is shot through with infinity."[86]

Thus the physicist and feminist theorist—a.k.a. new materialist—Karen Barad develops the relational ontology she has discerned in quantum entanglement.[87] Against the background of *Meeting the Universe Halfway* (2007), her great exposition of quantum indeterminacy and the "agential intra-activity" of electrons, she enters with Judith Butler the intertextual force field of Walter Benjamin. In her contribution to a transdisciplinary theological conversation, it is matter itself that implicates her in a theology vibrant with political significance. Highlighting the kabbalistic mysticism that influenced Benjamin's notion of the messianic now-time, she draws the following inference: "The messianic—the flashing up of the infinite, an infinity of other times within this time—is written into the very structure of matter-time-being itself."[88]

With no flicker of reductionism, such a cosmology undermines not just human but animal, and indeed organic, exceptionalism. The infinite is not an exception to temporality but the very infinity of times; the messianic does not take itself out of matter but inscribes itself at its core. At once mystically charged and scientifically exacting, this meditation channels the "weak messianic force" that challenges every top-down sovereignty. Barad's intra-active agency calls up an entangled multiplicity.

If, Barad continues, "matter has this messianic structure written into its finitude, *no matter how small a piece*, this is surely true of all material beings, each of which *is an enormous entangled multitude.*"[89] In the weakest, in the shortest of times and spaces, in "the least of these," is encoded an infinity of possible relations. To then link the political to the ecological is to underscore the precarity of this mattering moment—and also its indeterminate potentiality. It is to resist the temptation in every political emergency, however exceptional, to ignore the matter of the earth. The messianic inception "flashes up." Its mysterious infinity suggests a theology inimical with the sovereignty of the decisionist One and its Schmittian theopolitics.

A political theology of the earth, forged of entangled difference, calls upon that very multitude for Ecocene solidarity, which is to say, for our self-organization across vast reaches of critical difference. In their macrocosmic and microcosmic assemblages, the space-times of matter contract in and as *us*—just where we meet the universe "halfway"—here in the immediate entanglements of mesocosmic earth-time.

As to the politics of the matter, listen to what Barad wrote almost a year before the 2016 election; the physicist turns prophet: "And perhaps it is this eternal link among all living beings, all beings in their aliveness, this shared transience, and the possibilities for renewal that follow downfall, that is needed in confronting the rise of fascism in its connections with late-capitalism, the normativity of state-sanctioned violence against the oppressed, and the ongoing devastation of the planet and all its inhabitants."[90] In her remarkable summation of this earth moment, Barad mobilizes the possibility of renewal after collapse, Benjamin's "downfall," for precisely our earth moment. As with him, and so not without the political motive-force of theological metaphor, she resignifies the messianic structure as an "eternal link" among all living beings.

"Living" traverses the creaturely registers of organic and inorganic now constituting the animate ecology of the earthome and the schematic human threat thereto: "Facing the im/possibilities of living on a damaged planet, where it is impossible to tease apart political, economic, racist, colonialist, and natural sources of homelessness (otherwise called 'the problem of refugees'), will require multiple forms of collective praxis willing to risk interrupting the 'flow of progress.'"[91] These nonlinear multiplicities of resistance burst from the deep space of a materially entangled multitude, requiring an insistent plurality of practices. Channeling Benjamin, she adds: "not by bombing the other but by blasting open the continuum of history."[92] Each creaturely contraction of its collective throbs with timely possibility.

As Roberson opened this earth mediation between the small and large of time, so Benjamin cites a biologist on the immensity of prehuman life on earth: "the history of civilized [humanity] would fill one-fifth of the last second of the last hour." Benjamin comments: "The present, which as a model of Messianic time, comprises the entire history of humanity in an enormous abridgement, coincides exactly with the stature which the history of humanity has in the universe."[93] So the recapitulatory now-time is to human history as human history is to the unfathomable duration of cosmic history. No wonder the contraction of this fine-woven kairos may burst messianically through every schema of chronos.

The explosion that is dis/closure—prying open the impossible at the seam of its incalculable not-yet—invites not violence but breakthrough. Might we hear echoing in that blast Klein's awakening people's shock, the possible opening of an actually better homeworld? If the creaturely collective collects itself in vibrant, vibratory bits, bursts, births, might it not come into its own? Coming unowned?

The flashing up of the infinite takes place not in any predictable unfolding of chronos. But neither therefore does its kairos mark a final redemption. If the alternative comes "written into the structure of mattertimebeing itself," its planetary inception does not arrive as eschatological exception. This kaleidoscopic moment of a temporality contracting and opening across its divergent scales, always differently, performs an indeterminacy written into its undercosmos. Which has us surrounded.

Barad had already gone about "queering the quantum."[94] In the same stretch of time the whole universe has refreshed its mystery for this millennium's science as "multiverse," "dark matter," "dark energy." So why not inhabit already another schema—that of a universanimal that un-White-Outs time, that unwrites the straight times of progress or disaster, of redemption or The End? The Ecocene earth, amidst the closedowns, summons our collusion. Now.

Roberson returns in another poem to the subtle timing of our planetarity: *"Earth goes beyond us / Is the ours of cosmos / Is our hour of cosmos."*[95] In the contraction of the universe in "our" tiny terrestrial grain of it, the we-time of accounting has arrived. How will we live this hour of ours, this beyond, this unspeakable anthropocapitalochthulocenic now-time? Within the bigger hour of the species, now estimated at 195,000 years, wouldn't it be a marvel to round that up to, say, 200,000, when the species might die of a riper old age? Or, whenever, at least by some honorable death?

What matters are the possibilities breathing in us now. Our hour of cosmos. For now, we might just stay with the convulsive complications, remain with the multiscalar mourning, sprout at the limit, embrace the sensuous intersectionalities. We might assemble ourselves, our inceptional selves—in the midst of meltdowns—in the amorous agonism of earth's collective.

3

THEOLOGY

"Unknow Better Now"

Joined in a sovereign stranglehold, politics and economics grip the present of the collective future. So can the subtle now-time of the kairos not *now* twist open its possibility? The contraction that ingathers even as time shortens: a theological clue snatched from an ancient letter delivered a darkened hope. But if time is collapsing for the hospitable earth, how much time can *theology* have? Has "the Christ-event," as Clayton Crockett says, "run its course"?[1] Run out of time, out of breath, out of credibility for assembling the messianic potentiality of its progressive public? This is no mere white Christian crisis. Global monotheism is living off "the remains of God's corpse."[2] Oh Lord, RIP.

Such a judgment rings all the more disturbing as it is offered not in the service of a secularizing reduction but of a "radical political theology." Its alternative to the Schmittian brand of political theology, indeed to any strongman supplements to capitalism, fuels a deconstruction of Christian exceptionalism. We heard from Crockett that "theology is energy." To decode that energy he prefers Catherine Malabou's "plasticity" to the Christ-tinged force field of even the Derridean "messianicity." Such elasticity characterizes the biological symbiogenesis of Margulis and Haraway. And it resonates with the "possibilities for renewal that follow downfall" in all beings, as we observed with Barad and her electrons. Yet these possibilities mark the latter's engagement of an expressly messianic and as such political Judaism, evoking an "eternal link among all living beings."

If such theological plasticity energizes a now-time, we are only recapitulating the inextricably—but often unnameably—messianic kairos. In the present experiment, the very contraction of time, of present time and its traumas, is being read within the forcfield of a messiah that translates also as *christos*. But the question insists: has christology, however counterexceptionally conceived, run out of time?

To confirm the judgement of Christianity's exhaustion, at least in the global North, one need only google the latest Pew poll on "the changing religious landscape" to follow the steady, generationally cadenced, race-crossing dropping of the percentage of self-identified Christians in the U.S.[3] This is accompanied by the floundering or foundering of the old institutions of Christianity. (This is not just statistical—my friends and students work in them.) I had a poignant experience recently in the Netherlands, lecturing at the invitation of a theological organization.[4] The conversation was vibrant, located in a medieval church sanctuary stripped by the Reformation to a beautiful austerity. Enjoying the hospitality afterward, I was startled to hear from clergy and scholars that they are creating such new formats for theological conversation by way of coming to peace with the fact that the churches will close within one generation.

Most of the global North is ahead of the U.S. on this trend. We seem to be catching up. Each new generation comes less conditioned to the unconditional itself. And the distinct materializations of theos and its logos matter even and ever less. Good riddance, some mutter, with an amen from all progressives who place their faith, itself fading, in the progress of secularism. Good grief, others of us moan. We must let go of much to keep the grief good. But to mourn well requires a certain remaining—a "grave attending" (Karen Bray).[5] We stay with the troubles of theology.

The ghost is not one. Without melodrama we recognize the manifest failings of political, environmental, and religious institutions: a triple specter of apocalypse. I blink again—and it has still not gone away. This or that collapse, dissipation, extinction proves not to be the exception. Exhaustion gives way to failure. Of course the particular fragility of theological schematisms may seem minor in scale when compared to that of our ecosocial *world*. As political and as earthen, however, this theology registers its own failing as entangled in that of democracy and ecology. Amidst the contractions of this triune failure might theology itself perform something other than simple closure? Something opening into—failing into—a complicated darkness?

FAILING GOD

If theology fails, would that not mean that "God" fails? God. The very idea. And isn't the failure of God what is always at issue? Not merely theos as a rational hypothesis failing to be persuasive, let alone provable; but God, as so endlessly hoped and desperately prayed to, failing to come through for a person or a people? This failure presents early in Job's complaint, and ever since, as the crisis of theodicy, among assorted radical intellectuals, then, as the death of God, and now in the deadly statistics.[6] So do the death of God and the death of world arrive together? Or is that the death of *His* world? In the meantime, shouldn't I admit that this political theology of the earth is just another exercise in academic futility?

Please take none of these questions as merely rhetorical. However, we noted that the so-called end of the world begs the question of *which* construct, which schema, of "world." By the same logic, the so-called death of God begs the question: *Which* God? Which theological construction?[7]

The sovereign Lord who reigns through the Son over His creation has long appeared, at least outside His own unquestioning enclaves, to be failing to come through, to come, to *be*. But amidst our failure to embrace the Ecocene, His failure can no longer be masked by doctrinal evasions, feminist translations, and hopeful postponements. The failing God shadows the failing world, into which schematism He got (quite successfully, masculinity and all) secularized. The time of "His creation," set by biblical literalists at around six thousand years, bears no resemblance to the long times of the planet or even of the anthropos. But it does map loosely onto the scale of the geological Holocene. The biblical narrative of collapse from Eden into exile may of course soon read as a mere snakey foreshadowing of the coming fall: the ultimate exile from the garden, brought on once again by sinful ancestors. With their theological pretexts.

So the familiar world of the Holocene may indeed be accompanied in its collapse by its now globally familiar Creator. And so of course by his exceptional Son: *that* Messiah, *that* christology . . .

Even the collapse of a (relatively) long, stable geosystem does not however terminate the life of the earth or, necessarily, our life destabilized within it. It does highlight spectacular human failure. And unknowing. (I write while texting with my niece in Houston, amidst "unprecedented flooding"—an increasingly familiar kind of unknowing. With no hope that this season's ecotraumas will cut through the sovereign denialism.) So does this correlation of ecopolitical with theological collapse necessarily terminate the God-tangled kairos? And its Christ-carrying messianicity?

Or might they together entangle us in a livelier uncertainty?

If we experimentally persist in a political (OK, go ahead and call it Paulitical) theology of the ingathering earth, then we may consider this much: the threefold contracting of time does not exhaust the earth as such, nor, by the same logic, a possible theos of the earth. No more can the political reassemblage of that earth be ruled out—except where doomsday certitude rules. However, the triune crisis is, I have argued, exhausting the anthropic exceptionalism of our interlocked schemes for politics, earth, and God. Not only does the capitalobscene carbon machine continue to burn up our breathable future. But those who most suffer the entangled race-sex-class-immigration-health-climate depredations verge on an abyss of despair. Because of their ecoracial vulnerabilities, or because of their ecopolitical assessments, many surrender to a *nihil* of their own exhaustion.

Perhaps this is why theology comes yet again, here and there, into play. At this third fold of world crisis, its theos remains itself triply readable—but not therefore knowable. In its ecodivine all-contracting-all, with its logos of material intercarnation, its spiriting pneuma may be breathing beyond exhaustion.

THE WHITE VOID OF GOD

In the meantime, the political theology of the omnipotent exception flails violently. If Crockett observes His believers already feeding on His corpse, he is not making a point about the Eucharist. Rather, he infers that belief in "a rational, benevolent and omnipotent deity has become incredible."[8] On the Christian right, it "has been replaced by reactionary forms of evangelicalism and fundamentalism." The affective intensity of the reaction then at once conceals and compensates for the dead God.

This view gets confirmation in the enthusiasm of Christian conservatives for a president whose life exhibits one long exception to their moral code, let alone to the gospel's. Does he make up in charismatic antagonism what he lacks in Christ? He quickly began rehearsing an apocalyptic exceptionalism of "fire and fury like the world has never seen."[9] The phrase soon titled a best-selling exposé. But such rhetoric further consolidated the "we vs. they" of the conservative base. One does not know what enmity— nearer or further east?—may be punched up to Armageddon status.

In other words, the upgrade of sovereign emergency is only to be expected at the political interface of the triple apocalypse. Not that the religiopolitics of the right wants the end of the world just now—not when they command big button power. The end gets deferred, as we noted earlier, by what Schmitt, citing Paul, referred to as "the katechon," the imperial force that pushes back the threat of chaos. But recall also that the Bible contains no "end of the world." Its ancient imaginary anticipates catastrophic ecosocial collapses, not mere closure—imminent or deferred. The apocalypse dis/closes, unveils, the traumatic *failure* of the imperial world schema, not its katechontic good. This is just a way of saying—as *Apocalypse Now and Then* tried to in another moment—that closure might offer a neat and final solution.[10] But the Bible gives no excuse for it: no planned obsolescence of the creation.

The triple specter of a failing world schema, however—planet and polis ruled by the failing God—will not cease to haunt any emergent public. If we do not just shoo it away, it may help us to read certain blasts of *secularized apocalypse* more revealingly, which is to say, relevantly.

For instance, there fell into my hands this burst of apocalyptic intertext, published on the day of the 2017 inauguration. The novelist Pankaj Mishra writes: "Today, as white supremacists prepare to occupy the house built by slaves in Washington, D.C., it may be hard to resist the fear that these pugnacious men, 'struggling to hold on to what they have stolen from their captives,' as James Baldwin put it in 1967, 'and unable to look into their mirror,' will precipitate a chaos throughout the world that, if it does not bring life on this planet to an end, will bring about a racial war such as the world has never seen.'"[11]

Against the background of race struggles and colonial grabs, Baldwin's *The Fire Next Time* encodes a biblical warning, not a boast. The title contrasts to the catastrophe of Noah's flood the fire of Revelation. He will have had in mind not climate change but the nuclear threat. Now both are on offer. Mishra continues:

"Certainly, genuine democratic equality under the Trump administration will be a more formidable challenge than ever before. But at least it won't appear *veiled* by the illusions of the past—which may give present and future generations a better chance of bending the intractable arc of the moral universe to justice."[12]. Against the world-wrecking version of a democracy unwilling to face the white supremacism of its sovereignty, Mishra positions another image from the late sixties: King's messianic arc of justice. Not an arc bending on its own, but *getting bent by public action.* This hope has nothing to do with optimism and its illusory assurances. It anticipates—in exact translation of *apocalyptein*—an *unveiling.* The political eschatology here is faithful in its secularization.

Our fraught moment discloses a giddy and resentful grab-all, fomenting emergencies that can be blamed on the dark and alien chaos. Its sovereign decider bends not the intractable arc but the all too tractable rules of democracy. At the same time, there rips open already the alternate possibility—of the unveiling of the illusions of U.S. democratic egalitarianism. Then *apokalyptein* can be ecopoetically transliterated by way of Roberson's "un-White-Out" of language as dis/closure of a "better chance." That unveiled possibility comes "draped in black."

We followed in chapter 1 Kelly Brown Douglas' analysis of the unhealed history of Anglo-Saxon exceptionalism within U.S. whiteness.[13] It now seems that racial supremacism, hooked into the reproductive heteronormativity of antichoice politics, seals the deal with the Christian right. In the long-overt antiabortion rhetoric coupled with long-veiled racism is "unveiled" a remarkable secularreligious phenomenon: the culmination of biblical literalism in two religiopolitical causes with, literally speaking, no biblical mandate. Such theologically empty antagonism apparently serves to distract this Christianity from its own growing incredibility to itself. Baldwin's voice echoes in this white void from its half-century past:

> We human beings now have the power to exterminate ourselves. . . . We have taken this journey and arrived at this place in God's name. This is the best that God (the white God) can do. If that is so, then it is time to replace Him—replace Him with what? And this void, this despair, this torment is felt everywhere in the West, from the streets of Stockholm to the churches of New Orleans and the sidewalks of Harlem.[14]

Baldwin does not whiteout God. He is answering the question—*which* God?

The white One, at our point in "this journey," gives new evidence of His failure. Vis-à-vis our progress toward self-extermination, we had considered another unveiling, the one that can unleash Naomi Klein's promising People's Shock: that of the connection of global capitalism to climate catastrophe, the extermination slower in timing than the nuclear option. If this chain of supremacisms—racist, masculinist, anthropic—reads as secularization of the old Christian exceptionalism, one can honorably go for godlessness. But then does one invest one's faith in a new—progressive, better, even revolutionary and messiah-free—exception? Whiting-out God after all?

Alternatively: we stay with the troubles of theology. And so with our god-tangled historical present.

PALE OPTIMISM

With Benjamin's Angel of History, our gaze falls on an entire past unveiled, ethically pressuring the present. But in the present contraction, the gaze cannot be limited to the past. The Angelus Novus may be staring at a not-yet that is still indiscernible from its unrealized past. The "thick now" is thick precisely with a past that painfully contracts and presses toward birth. There is nothing sentimental in its messianic returns. A renowned verse of dark poetic response to the First World War, for instance, offered no sweet babe for the Second Coming: "And what rough beast, its hour come round at last / Slouches toward Bethlehem to be born?" (Yeats). The brutal Beast of the Apocalypse or the challenge of an earthbound divinanimality?

In the unveilings there flashes up the prophetic eschatology that was contracted of Hebrew experiences of desperate migration, of royal betrayal, of involuntary diaspora. The prophetic voice emerged in Israel less as a prediction of pleasing outcomes than as confrontation with the injustice of the sovereign, even already with Nathan rebuking the charismatic King David. Through multiple failed protests against subsequent state powers, prophets warn of dire consequences for the land, human and nonhuman, if infidelity to the covenant of justice continues. In its biblical origins, hope cannot be confused with optimism. It was always agonized and agonistic, confronting critical difference, mourning unbearable loss, and yet struggling for the new— the polis of a new theopolitics, New Jerusalem, new heaven and earth, new creation. "New" as hope was itself quite new, not a high value in other civilizations, which trusted more to cycles of return. The *novum* signified the gift of a promise, the promise of a gift. But the promise is not the guarantee, and the new is not an advertisement. The gift, as Derrida reinvigorated it, is unconditional; not given in expectation of a payback, it gives the very conditions of responsibility, of free response.[15]

Love knows this grace, this present, of inception: of the presenting, the vividly present, possibility. And hope is what carries gift into future, beginning with that expectant opening into the next moment. It holds the new as a promissory possibility. Hope, as we read with Bloch, is historically the condition of militant materialization and of creative embodiment. It was in Christianity (and in some readings of the gift) too readily forgotten that fulfillment of the promise depends upon a covenantal reciprocity—the indeterminacy of a people doing our part.[16] So the hooking of hope into certitude, as teleological expectation of a triumphant outcome (Christian or secular) has worked to trump hope itself. Then hope turns into what Lauren Berlant calls "cruel optimism," which she defines as "the condition of maintaining an attachment to a significantly problematic object."[17] Such pale optimism locks its believers into investments in a future that will almost certainly fail the present possibility. Hope lives—if it enlivens rather than deceives—not in predictions of "end things," but on the vibrating edge, the *eschatos*, of a precarious present becoming.

This edgier becoming has often taken nontheist form: "Revolution is absolute deterritorialization even to the point where this calls for a new earth, a new people."[18] The Deleuzian deterritorialization of terra itself unlocks the energies of our cosmic matter (as a "cosmic earth") for a new becoming, freed from the stranglehold of capitalism and its democracy. New earth, new people; one could, for purposes of a radical political theology—of the earth indeed—let this sheerly immanent phrase replace the ancient eschatology, freeing the "new" of Christian cliché. I would for a while. But, without the unmistakable echo of the ancient "new heavens, new earth," would the secular translation not quickly pale in tone? Instead of a secularizing supersession, why not contract, fold together, the ancient prophetic hope—so unwhite in its origins and hugely so in current iterations—with the creative force of continental philosophy's "repetition with a difference"?[19] If for Deleuze difference contracts as repetition, a political theology of the becoming earth contracts its prophetic antiquity in the difference of a secularreligious commons.

CHAOTIC EDGE OF CREATION

Let us then consider an *eschatology of the creation commons* as part of a recalibration of the whole discursive spectrum of theology. It welcomes the radical immanence that corrects all supernatural exceptionalism. Pure immanence, the domain of radical political theology, may better come into its own, however, not as an erasure of transcendence but in the humbler origin of *transcendere* as "climbing across"—not a flight from but a *struggle through*. This transcending does not take us out of the world but across it: into its own beyond, the beyond within, the critical but not separable difference. We "climb across" one world schema into another possibility, which may be known only as not-yet-known. It will take us into rough terrain.

Such an eschatology does not await a final sovereign decision from above, gradualist or apocalyptic in its imposition. The "new creation" does not signify, even in the biblical apocalypse, a final solution but the dis/closure of the participatory collective of the creation in its interspecies, intraplanetary agency. And that means the awakening of human responsibility as the more than ethical, ancestrally messianic, creativity. It spurts from the heart of the creation, which begins always again now. Doctrines of creation and eschatology then form not two ends of a time line but the alpha and omega of any moment of becoming (genesis).

In order to mobilize the force of the creation, its capacity for novel materializations, it remains tempting for any political theology of the left to appeal to a *novo creatio ex nihilo*.[20] This has its liberative theological representatives, its faithful messianism. But such a miraculous narrative marks the creation as exceptional origin driving time toward an all the more exceptional goal. Then transcendence threatens to abandon the amorous agonism of an actually transformative now-time. The newness out of nothing thus evades the messier eschaton of the edge of chaos. And so it does not face up to the failures of its God, for it carries the old habit of His guarantee: If God doesn't come through now, He will in the End. Secularized as progress or as revolution, the confusion of hope with certainty is certain to disappoint.

So then to unleash the cocreativity of the mottled creatures of the creation commons, we construe eschatology precisely not as the temporal exception, doctrinally speaking, to the ongoing creation; we are speaking instead of a *novo creatio ex profundis*.²¹ The creation commons reads out as the eschatology of the *creative* commons. The churning deep keeps us on edge. The chaos does not quit. On that eschatos we may tense up in defense. Or we keep learning to breathe with its pulsing spirit, to stay with its contractions.

THEODICY: HOW COULD HE?

Still, isn't such a messily multiple and uncertain outcome all the more disappointing? Theists have every right to wonder if this doesn't just leave us with a miscarrying status quo, an impotent immanence. What new can come of that? From some God of any or all colors that can only fail our hopes?

Yet thoughtful theists recognize that "God" has with numbing consistency failed to show up in the standard accoutrements of sovereign power. Or, more precisely, of omnipotence still yoked to an ideal of the good. But a *deus omnipotens* whose might suspends and transcends any humanly recognizable good or right—a particularly Calvinist solution—can by definition not fail. Whatever happens does so, and by definition, at His will.²² That is a long-time solution to the quandary of theodicy: God so strong, He can't be wrong. Hence the version of unacknowledged political secularization so effectual for the religiopolitical right: raw power freed from ethics. Feels divine. But then when the wrong hits home it fails our expectations for His intervention. That is the old news of theodicy: if this suffering, this trauma to my person or my people, is the means, what end can possibly justify it?

If any events that occur can be attributed to the omnipotent perpetrator; it is then merely justice that fails—or rather is deferred until the End. In the meantime, it's a mystery. A meaningful apophasis is thus chained to an all-too-knowing sense of all-controlling power. Hence a certain voluntarism, an arbitrary decisiveness, always inheres in the doctrine of omnipotence. This, we noted, is the political ex nihilo of the Schmittian sovereign. An indiscernibility between outcomes worldly or otherworldly offers theological cover. It protects theologically and politically against the purported rationalism of the perennially anguished question: how could a good God let . . . ?

What has contributed more to the dwindling of mainline Christianity and indeed Judaism (publics relatively educated in critical thinking) than the perpetual pastoral crisis of theodicy?[23] *If* God could have prevented but "let" this horror, this child's painful death, this ancestor's enslavement, this people's holocaust, happen, for His own inscrutable reasons—to teach, to punish, to test—*then* atheism is the only answer. With Dostoyevsky's Ivan, I throw myself into the crowd who can only "hasten to return my entrance ticket" to heaven.

Yet if I linger with the "if," and with the less hasty Karamazov, the meditative Alyosha, I wonder if the "the death of God" whites-out anything other than the Lord God of the sovereign takeout.

Eschatology takes us, at its own edge, right to the question of God's power and therefore of God's existence. For it is in some image of divine power that, with reason, the ideals of human leadership have been forged. So then a political theology of the earth can only work if it has a notion of divinity that has faced into the contradiction of the all-good and all-powerful God. If it does, it does not thereby need to convert to any alternative theos those who have given up on God. It does however hold open a cloudier, uncertain space, the space of an "if" at the edge of the "is." Really uncertain. A dark-out of determinacies.

QUEER ARTS

Would some God cloaked in amorous darkness now succeed where the traditional deity brightly fails? Or might theology withdraw from the competition? In other words, we might take a cue from what Jack Halberstam famously calls "the queer art of failure":

> There is something powerful in being wrong, in losing, in failing, and . . . all our failures combined might just be enough, if we practice them well, to bring down the winner. Let's leave success and its achievement to the Republicans, to the corporate managers of the world, to the winners of reality TV shows, to married couples, to SUV drivers.

Halberstam is not bothering with God (they have enough on their plate). They are talking back to the politico-capitalist scheme of success, which blames the inequities of its systems on the failure of its victims. We should not be shocked that Halberstam draws here upon Benjamin, with his revolutionary but antiteleological refusal of progress-temporality. "The concept of practicing failure perhaps prompts us to discover our inner dweeb, to be underachievers, to fall short, to get distracted, to take a detour, to find a limit, to lose our way, to forget, to avoid mastery, and, with Walter Benjamin, to recognize that "empathy with the victor" invariably benefits the rulers."[24]

Success seals the privilege of the winners, the victors—the exceptions. And it is a seal imprinted on its few, including, of course, select "classy" exceptions of color and sex, with the white force of heteronormativity. Capital haunts our every "climb across" to critical difference. But losing, wandering, and falling short take us off the straight and narrow path of success. Then we may fester with resentment. Or stay with the trouble of the uncertain alternative.

Similarly, in a theodicy in which the victims carry the blame for their suffering, shame due to private "sin" conveniently occludes suffering due to social oppression. It teaches identification upward, with those apparently more favored by God. So then protest against the structural injustices of race, sex, and class will intensify the resentment of those trying to stay on the straight path of success. Spirited alternatives will be repeatedly doomed to failure. Failure, however, may provoke more creative, generative, and surprising ways of inhabiting our world: "Rather than resisting endings and limits, let us instead revel in and cleave to all of our own inevitable fantastic failures."[25] What might such reveling reveal not only about theologically shaped lives, but about the theological shape of "God"?

Rather than just transcending God and succeeding at secularism, or, to the contrary, judging the secular to have failed and winning God back through the postsecular, what if theology keeps faith with its own failures? This does not require that we repeat them. We would become alert to our own ploys of power and certitude, our own sovereign subjectivities. We would practice mindfulness of the complex affects of failure. We would watch for possibilities buried in the ruins. We would have the benefit of a half-century of experimentation in a queerer art of theology, in the multicolored she/he/it of divinity, the LGBTQ-they mutability of what we nickname God, and so also in the plurisingular, pancreaturely *we* of the imago dei.

In its contemplative evolution, this politically charged agonism of images minds its own metonymic proliferation. Its multiplicity expresses a creativity undaunted by the very failures that release it. If normative sex/gender roles fail, the queer arts do not then "transcend" sexuality—except in the root sense of clambering across tricky terrains. Adventurous revelry stumbles across discarded revelations. For the shifting landscapes are littered with theological ruins.

An Yountae therefore explores "mysticism and cosmopolitics from the ruins." This scholar of religion, like Crockett, turns to Malabou and the plasticity of being. In his meditation on the ruins—of a colonial (white) God, His world, and His shifting, creolized subjects—An leaves us with this clue: "It is perhaps here that the boundary between the human and the divine or between the spiritual and the political dissolves (at this juncture of the groundless middle)."[26] So it seems the secularreligious comes tuned to an abyss. Its boundary-crossing fluency comes steeped in the fluids of the deep. Its emergent public *remains* in its intersections, artfully minding the precarity of cosmos, of politics, and of theology itself.

BRILLIANT DARKNESS

In the subtle luminosity of a difficult darkness, a "polysemy of the abyss" becomes possible.[27] Here an ancient experiment in utterance mingles with such current ones as An's decolonial cosmopolitics. In the "luminous darkness" of Pseudo-Dionysius, then in the divine abyss that Eckhart called "the Unground," a groundless middle opens between theology and its object, one that refuses all objectification. Here any names of God—including *God*—go dark. To the question "Then how should I love God?" Eckhart responds climactically, "You should love him as he is a non-God, a nonspirit, a nonperson, a nonimage . . ."[28]

Negative theology, bound to its affirmative traditions, stays with the failure of theological language: which is to say of any finite signs trying to capture the infinite. As "infinity itself," the divine remains, as Cusa noted, a pure negative, being no being that has a boundary, an opposite, or any single name.[29] Inasmuch as it "is," it is discernible only in an indiscernible plenary, a foundationless deep, a cloud of impossibility, a semantic detour. "The truth is not merchandise."[30]

The apophatic alternative can never *succeed*, as it would in the next moment have to negate the terms of its own success, of its own utterance. This mindful failure of language is then reason not for a shamed silence among those of us who occasionally make a theological point. It is a stimulus for the courage of a creative plasticity. Yet it also, and at the same time, grieves the failures of theology to persuade, to motivate, to sustain, to comfort, and to provoke. It laments the arrogance of so much theology, failing to recognize its own failings and therefore perpetrating a political theology of sovereign certitude. It regrets the failure of nearly two millennia of the tradition of negative theology to redirect Christian mainstreams toward ethico-linguistic honesty. So the very recognition of the failure of theological language fails 1. fortunately, to remain silent; 2. unfortunately, to overcome the power plays of an unquestionable Word; and so 3, to reach the majority of theologians who suspect that the apophatic discipline means just, in Monty Python's parlance, "Oh, shut up." And yet negative theology has been munificent in a language of words edged, not erased, in silence—in the language of its glowing darkness.

Negative theology did not drive the modern secularization of God. But we may recognize the gesture of secularization as its own unsaying of God, known or unknown. A theology that *knows* its own unknowing reveals the abyss at the heart of all talk that matters. It may help us now to live amidst the ruins of religious and political certainty. So the breathing room that negative theology offers as its own affirmation—call that the space of the Holy Spirit, if it helps, the pneumatism of the abyss—offers a matteringly meditative renewal. It may spiritually recharge but not replace the political work of an amorous agonism.

Very practically, indeed politically, a margin of self-knowing unknowing empowers the encounter with critical difference. It smudges the binary opposites and allows for the "I hear you" of negotiation. This is true up close, as among clashing progressive priorities, or at the remove of an undecided public. An efflux of indeterminacy opens intra-action within uncertainty. Some groundless middle, more fluid than orthodoxy or compromise, can dis/close itself. A collapse becomes a contraction. Consider the world-shifting effects, as, for example, in the movements of Gandhi and King, of old contemplative practices, Hindu and Quaker, of silence. Did those movements "succeed"? Surely they achieved no final triumph, given the assassinations, betrayals, disappointments, the backlashing generations. At the same time, however, they came closer to realizing their own hopes than have most political events of assemblage for justice.

In an eschatology of the creation commons, the messianic potentiality is carried discontinuously into the future, in failures and realizations, through multiple successor movements of nonviolent agitation, as the unrealized potential of the present's past.

"EVER TRIED. EVER FAILED. NO MATTER. TRY AGAIN. FAIL AGAIN. FAIL BETTER." (SAMUEL BECKETT)

To draw upon failed theological histories, with their moments of kairotic breakthrough and their cruciform disappointments, does not inspire optimism. But in-spire it can, delivering the breath of a stranger spirit. Halberstam's queer art of failure is, after all, not a nihilism of antisuccess and nonachievement. It is "about alternative ways of knowing and being that are not unduly optimistic, but nor are they mired in nihilistic critical dead ends. It is a book about failing well, failing often, and learning, in the words of Samuel Beckett, how to fail better."[31]

The allusion is to *Worstward Ho*, one of the last writings of Beckett. (The title alone takes out U.S. exceptionalism and its westward progress narrative.) Beckett always performs language at the edge of itself, at the eschatos of an utter indeterminacy. This language on edge, failing to be captured by the signifier, failing (white guy un-Whiting-Out English) to capture the Transcendental Signified, wreaks havoc with sovereign grammar. So then it breaks into revelatory chaos:

A place. Where none. A time when try see. Try say. How small. How vast. How if not boundless bounded. Whence the dim. Not now. Know better now. Unknow better now. Know only no out of. No knowing how know only no out of. Into only. Hence another.

A disoriented now of space-time, micro and macro, contracts into a language of apophatic unknowing: the intra-activity of now and know, each negating itself, negating each other, only to "know better now." Which is to "unknow better now." Not to lay God on Beckett, either—but maybe (waiting for Godot, God-out) he had encountered some version of negative theology. Knowing "only no out of": no escape, no extraction from the universe, no mystery of takeout transcendence, it performs its own irreligious mysticism. In the loss of the grammatical "I," the pronominal anchor, there occurs movement beyond—yet "into only." No exit, no exceptings. "Hence another." The other, the difference, the relation always already immanent. Not taken out.

Earlier, and otherwise, Whitehead had named such inextricable difference "mutual immanence."[32] He was marking—across every living being, every being as alive—a "no out of." It collects us in our othernesses, know it or not, like it or not, inextricably. Each creature becomes, in its moment; and it composes itself, for better or worse, out of its world of entanglements. In this way we might say that the knowing better, which is an unknowing better, performs the epistemological corollary to a cosmology of failing better. And that better failure—which does not in itself distinguish humans from other failing and evolving species—carries the whole world on its shoulders. Poor Atlas.

Some cosmologies will fail better than others. This will have to do with their recognizing their own "first principles" as "metaphors mutely appealing for an intuitive leap."[33] In the cracks of cosmology, between its world-words, the queer art comes—if it can—into its *theological* own. Might we, unknowing better, knowing better that we do not know, also fail better in this time of theological downfall? Might we let God—Him/Her/It/They/Thou/We/Other—fail better? The straight Word of the He-God is let to fall into its own confusions. Internal contradictions, as between omnipotence and goodness, drive it into the dark cloud. *Whence the dim.*

His failures to fix the world then lose the power to surprise and disappoint. To demand scapegoats. The success or failure of that God at defeating evil or managing the good(s)—let alone at keeping His institutions afloat—ceases to be the point. Even the bottom line of God's "existence," in which you "believe," yes or no, falls beside the point. It falls, in fact, beside the corpse. *His* death multiplies in its decomposition. But then the cosmic life we nickname God would no longer signify the maker or breaker of success, the One who has failed so systemically to fulfill His promises. Does there ripple in His wake an abyss of possibility?

"How if not bounded boundless." Here, sparks of momentary alternatives fly. Here, God most certainly fails to meet our expectations, theist or atheist, of a controlling care. S/he/it might never have meant to *succeed* so much as to *materialize*. What if the unconditional *matters*? Among all manner of material conditions? If so, this embodiment would to an alarming degree depend upon the responsive bodies of creatures. The signifiers of the imago dei, of the incarnation, of the bodily resurrection of all, of Meister Eckhart's extension of the nativity to each birth of God in the soul[34]—these are rich old codes of divine *mattering* in the world, a materialization happening not through control but through collaboration. *Creatio cooperativa*. It would be less a matter of a plan being fulfilled than of experiments being enacted.

On earth the experimentation seems to be taking place across multiple temporalities, through endless turbulences and layers, trials and failures. There is no evidence that it will or must "succeed" in some lump sum of final judgment. There is some evidence that it may cut its own experiment tragically short. Whatever: endlessly much is creatively generated, lived, lost, enjoyed, and suffered along the way. A person may hope to grow into a sense of "tragic beauty."[35] What we might still on occasion call God offers neither a dictation of the plotline nor a dictatorship of ends. God, in the process schema, offers the lure of novel possibility. Amidst no matter what failings. Not, we shall see, as the miracle of exception to the world, but rather as the grace of inception within it.

EROS OF THE UNIVERSE

If God has failed to come through, we have always already failed God. It may be that "we"—in our systems of abstraction and extraction—have thwarted our collective possibility. Perhaps "distraction from distraction by distraction" (Eliot) is blocking its realization. Realization signifies mindful actualization. It does not mean our *belief* in God, who, if the trope has any life in it, could care less if you "believe in" Him, Her, It, They, Pan, or Nihil. In this, divinity remains rather Zen: enough already with the "Lord, Lord." Cut through illusion and reduce suffering![36] What matters is what possibilities we cocreatively, each and all, the Genesis collective, are realizing, are mindfully materializing. Across our most critical differences. Not what we are finally achieving, but what we might be—in each contracted now-time—nursing, fomenting, ingathering.

In process theological language, God folds into any moment, into any creature, now, as "primordial nature," a lure, a nuance of possibility, and, on the other side of that moment, as "consequent nature," takes in, feels, suffers, *em-pathos*, what the creature has actualized. Lure and affect: no control.[37] So far from the Aristotelian/Thomist metaphysics of the Unmoved Mover is this divinity, this "Most Moved Mover" (Hartshorne), that the divine movement in the world simply cannot be likened to that of the movers and shakers, the sovereign success stories of history.[38] This movement rather resembles, if it can be said to resemble, someone who loves, loses, and then, like brother Alyosha, gives amorously more.

This theos, however, is almost all nonhuman, as is its world. In that boundless body of energy and matter, mostly dark to us, the earth contracts the world into a little galactic organ. Yet the space-time of the cosmos—with its "spooky action at a distance" (Einstein)[39] materializing hints of instant interconnection across any expanse—is not envisioned as itself indifferent. Each thing feels its relations as it becomes. Recall that with Cusa its boundless cosmos contracts to every creature, in its precise and bounded *difference*. In the illimitable relativity the "shaped togetherness of things emerges" (Whitehead).[40] The things are called forth in their distinctness, each edged with possibility. But they cannot escape each other: "Actuality is through and through togetherness."[41] (*No out of. Hence another.*) And "God" names here a principle of concretion, provoking new particularizations, new differences, new shapes of bodied togetherness. As its own.

As John Cobb has christologically unfolded Whitehead, the divine is always everywhere therefore soliciting embodiment, incarnation.[42] And, at least among process theologians, failing better than most theological movements has solicited a shaped togetherness of planetary ecopolitical materialization. So talk of the incarnation remains for a political theology of the earth meaningful to the extent that it is not signifying one exceptional Son but rather a distinctive breakthrough of something happening, in potential, everywhere—call it *theopoiesis*, call it the becoming all-in-all, call it intercarnation. But take responsibility for *your* calling.

Upon learning of a systematic alternative to classical omnipotence, most theists do not congratulate us on our queer art. They might respond instead: "Isn't this just immanence and impotence filling the void of sovereign transcendence? Weak! No wonder liberal Christianity is failing." By the same token, this subtler theos, as "Eros of the Universe" (Whitehead), as ecogod or as revolutionary love, annoys atheists, who want the All-Powerful Him as worthy adversary.

So the embrace of that very weakness takes place in the secularreligious third space, groundless middle between straight secularism and even straighter orthodoxy. Outside of process theology, its impactful text is *The Weakness of God*, in which John Caputo mobilizes deconstruction for an ineluctably theological project. He had already offered a magisterial rendition of Derrida's theological reverberations along the two registers, the messianic and the apophatic, that, as we have seen, converge upon any political theology liberated from its deus omnipotens.[43] Caputo's philosophical method, far from Whiteheadian, yet similarly cuts away the substance metaphysics of an unmoved Mover and its sovereign certaintism. His "poetics of the impossible" vocalizes, against omnipotence, "the weak force of the call."[44] For Caputo, this weakness—framed as "after" but not *as* the death of God—expresses at once the deconstruction of God's existence and the surprising experience of God's insistence. Echoing Richard Kearney's *God Who May Be*, Caputo proclaims a God of "perhaps." Kearney's anatheism—God happening again after "the death of God"—itself channels Cusa's God as posse ipsum, "possibility itself." If God "fails" in this postmodern iteration, it is an artful happening.[45]

In Caputo's succeeding volume: "God does not exist but insists." But then also: "God insists upon existing."[46] One may be left wondering whether *that* insistence is ever honored. If so, then God must after all in some sense consequently exist. Does this possible oxymoron between existence and insistence yield another *coincidentia oppositorum* teetering at the edge of language? One might infer that God, *s'il y en a*, only has existence, actual life, and movement in and through creaturely events: like Paul's messianic fall on the road to Damascus, like Eckhart's apophatically blooming rose, like the Lakeland high school students' uprising, like . . .[47]

We could experimentally imagine that God's existence happens only through the entangled materializations of the world: the "sensuous intersectionalities" of a promiscuously distributed intercarnation. Some More, minimal or immense, insists—but only as possibility. The possibility can along with "God" remain abstract, or it can contract with a history and so become existent. (*Hence another.*) In the contraction of "what has been and what might yet be" (Eliot), the actualization comes entangled in the world it recapitulates. This would not then ground God's existence as an established Being. But that existing might flash up. In the now-time.

Then the call, the gift of a chance, the lure to try again no matter what, becomes all the more insistent. All the more cosmopolitical. Its divine ecology discloses itself as dependent upon our agency, our humanimal-plant-geo-quantum intra-activity. It is never just our human or political or even earthen existence that is at stake. So it is the deconstruction of a sovereign Lord that relocates the creator/creature relation within a political theology of the earth. And that entails not a destruction of (that) God, not an incineration of His corpse, not our own exceptionalist dissociation from that failing Father. Majestic traces linger, darkened, disarmed, disarming.

WEAK MESSIANIC POWER

So then it must be said that in terms of the paulitical, the primary text of "the weakness of God" is none other than 1 Corinthians: "For the foolishness of God is wiser than human wisdom, and the weakness of God is stronger than human strength" (1:25). Caputo rightly relinks this locus classicus of divine weakness to

Paul's sense of the inbreaking kingdom of God, the *basiliea* that flies in the face of the sovereign empires.[48] Paul is writing to the Corinthians about their own apparent powerlessness and their lack of accredited expertise. His point is this: "Consider your own calling, brothers and sisters: not many of you were wise by human stndards, not many were powerful, not many were of noble birth" (1 Cor. 1:25f). In other words, the lure calls, it insists itself (in his context) across class, race, gender, marital status, ability, educational level, and religion—across critical difference (in any context). It is what is calling to your becoming, uniquely and together, "members of each other" (Rom. 1.25) now.

The weakness of God is embodied for Paul first of all in the cross: "The messiah, the power of God" is manifest as "the messiah crucified." Not as glorification of suffering and perpetuation of systems of victimization—as the crucifix was so systemically deployed—but to the contrary, as an empowerment of the weak: "God chose the weak things of the world to shame the strong" (1 Cor. 1:27). And then, in a gesture cutting radically against the whole hierarchy of being, "God chose the lowly things of this world and the despised things—and the things that are not—to nullify the things that are" (1 Cor. 1:28). Things that are: the substantial beings of the status quo. Things that are not: those lacking substance, property, proper being, who do not rate even as "beings" in the ontology of the empire. Language nullifies, convulses, inverts, goes dark at the edge of the speakable. The be/able.

As James Cone's great analogy of the lynching tree with the cross has it, divine solidarity with the descendants of slaves— "things that are not"—cuts all the way to crucifixion.[49] But then a horrific execution can no longer be written off as God's will. No theodicy can sanctify it as a divine act, like the penultimate act in a Passion play. But what drips and cuts through the narrative of crucifixion is a revelation hidden in plain orthodoxy, concealed in atonement theology: that the passion is truly a passion, *pascho*, suffering. Jesus' Abba does not cause but *feels* the torment; heresy to that Unmoved Mover, the passion reveals divine vulnerability.

If Luther first marked the cross with "the death of God," it was not to preach that God thus ceases to exist. The Lutheran Hegel developed that theologoumenon as giving way to the immanent becoming of the Spirit of history, and his heirs still bat around the meaning of God's death. But Luther was insisting, against classical orthodoxy, that—for that messianic moment at least—God *as living* enters full into human agony. "Luther observed that God is to be found precisely where theologians of glory are horrified to find him: as a kid in a crib, as a criminal on a cross, as a corpse in a crypt. God reveals himself by hiding himself right in the middle of human existence as it has been bent out of shape by the human fall."[50] To this existence, God ceases to afford the metaphysical exception. But that indwelling "in, with, and under" creation does not diminish divine mystery—*Deus absconditus* and *Deus revelatus*—but redistributes it.[51] The apophatic expresses not the distant exception to creaturely life but its grounding Unground.

The weakness of God implicated in the deconstructive "religion *sans* religion," in Derrida's messianicity minus messiah, has no truck with any metaphysical exceptionalism. Indeed it carries the shadowy intertextuality that we have glimpsed throughout this meditation: Walter Benjamin and his "weak messianic power," which Taubes and Agamben link in its complex Judaism to the no-less-complicated Judaism of Paul and his weak messiah. Taubes emphasizes that "theology is for Benjamin messianism/the Messiah. It seems to be implausible, a dream, a myth."[52] So the historical materialism that Benjamin advocates would then need to "enter into a pact with it" through the development of "a concept of theology that can go together with the radical immanence of historical materialism."[53] As an irruption immanent to history, the messianic comes as a momentary burst, an interruption not of time but of its transcendent teleology.

In a related vein, the Roman Catholic theologian Johannes Metz proposed a "so-called new political theology," new first of all as an explicit embrace of the phrase in opposition to Schmitt. Embracing secular critical theory, it revalorizes the Jewish roots of Christianity, in which the political was never split off from the spiritual, over against the predominant schematism. And for Metz it is *political* theology . . . insofar as it situated religious practices and religious reasoning in the given historical time and socio-political context of late twentieth century and early twenty-first century's globalized world. In fact, the new political theology could at least in part be read as a comment on the writings of Benjamin."[54]

As commentary on Benjamin continues to proliferate, new twists of timeliness flash into focus. Crucially, the messianic weak power never refers for him to a singular divine appearance—not in a past incarnation, an eternal present, or a destined future. Rather it is "the future of the past," that which, in recollecting (not resurrecting) contracts its tensed present. Here Judith Butler inhabits Benjamin: "The conclusion is not that the messianic belongs to another order, but only that it operates within this one as a constitutive alterity—breaking in, breaking out, flashing up, confounding without collapsing the spheres of the this-worldly and the otherworldly."[55]

HUNCHBACK THEOLOGY

In the last chapter's reflection on the "end of the world" vis-à-vis the multiple temporalities of a fierce and fragile planet, we were ineluctably drawn into Karen Barad's reading of Butler with Benjamin, which led her physics to display a messianic texture of "space-time-mattering" itself. Benjamin himself evocatively links the messianic to the temporal structure of the material—and not merely economic—cosmos. In so doing, he contextualizes the human histories of annihilation and suffering, of all manner of "downfall," within a vibrant "nature." It is pulsating beyond any anthropic exception: "The rhythm of this eternally transient worldly existence, transient in its totality, in its spatial but also its temporal totality, the rhythm of Messianic nature, is happiness."[56]

And then, no less startling: "For nature is Messianic, by virtue of its eternal and total passing away."[57] This totality of transience could suggest a messianism of apocalyptic closure. But then the passing away, the finitude, would not be "eternal." This messianic nature must be read in its character of "eternal transience," signifying the downfall of every life, of every bounded creature in the boundless rhythm of space-time. It reverberates with a notion unknown to it: that of his contemporary Whitehead's "perpetual perishing," the demise that marks—rhythmically indeed, as each becoming belongs to the "vibrating universe"—every moment of becoming.[58] No exception. And the very notion of "Messianic nature" undoes any exceptionalist messianism: its difference is not absolute but constitutive. It participates in that to which it is "another": entangled difference theologically energized.

Butler muses with Benjamin that "happiness may well be an internal musical dimension of mourning and sorrow, if not its outcome." (As, perhaps, we may experience radiant well-being listening to Mozart's *Requiem* or singing the blues. "Tragic beauty.") Butler then asks: "Can we understand happiness as the apprehension of the rhythm of transience and even consisting in the abandonment of an anthropocentric relation to loss?" This is a powerful question, surely a self-interrogation, from one preoccupied with human mourning, one who cannot in her own philosophizing readily let go of "the anthropocentric conceit."[59]

Staying with the troubles, minding human betrayals of the Ecocene, we may experience this rhythm ever more vividly. The translation of the messianic into a transient All of the world exposes a cosmological precarity, bigger than us, bigger than earth. It lends human depredations a context even as it denies them an excuse. In that "totality," the scales of time catch every historical downfall: hence Benjamin's "enormous abridgement."[60]

"How vast. How small" (Beckett). "That very subtlety of time between large and small" (Roberson). So this conversation offers a corrective to the anthropocentric middle to which any historical materialism, any messianism, and any political theology of human struggle will be tempted.

Of course, this rhythm of transience may also correct for theological certainty. It may at once expose our historic failings even as it enlists our lament and our laughter. The messianic counterexception, embodying that "nature" that births and obliterates us, may get a word in edgewise—an eschatologos, where every end is the precipice of possibility. The death of the One sovereign Lord then does not assure the demise of theology and the blockage of new secularizations. It does drain the grandeur from the project—the stony majest-He worn down to queerer geological formations.

In the "Theses" Benjamin had cast theology in the role of a "hunchbacked dwarf," to be simultaneously "enlisted" and "kept out of sight." As these days its hunch is growing deeper, given the failing of theological intuitions and institutions, it is getting a bit too easy to keep out of sight. Yet the strategy of an invisibility enlisted for secular work directly mimics that of political theology. So we are trying to keep theological effects—sovereign or messianic—at least in our sights. No pure and simple invisibility for this hunchback. Otherwise theology cannot be held responsible. Therefore the reminder of negative theology, which sees and so is seen only in the glow of darkness, keeps mattering. It rehearses us in the interplay of seeing and unseeing, knowing and unknowing, saying and unsaying. So it can help keep the interplay of visible public effects and hidden theology itself in sight. Theology in its cannier truth comes always already deeply bowed.

As its failing body hunches ever closer to the earth, its own ancient practice—in the name of mysticism, spirituality, religious plurality, or religiosecular alliance—may be enlisted for surprising new publics. That these will be ecosocial publics means that we know ourselves utterly dwarfed by the planetarity of the challenges ahead. As to the earth, the entanglement of the messianic in Benjamin's "nature" and in Barad's "quantum queerness"—that is, in the multiscalar spatiotemporality of the earth—energizes the political life of a theology that resists the anthropic exceptionalism of both theism and atheism. That life, as we have argued, depends upon the enlivening effect of the nonhuman. In an ecopolitical theology that nonhumanness always dims into an indiscernibility between divine and creaturely figures of animacy. As theology in its contraction pulls us down closer to the earth, is it beginning to tune to our animal kin, our creeping and crawling, rarely erect companions? And to our planted ones, moving in roots and rhizomes in and out of the ground of our being? Deep hunches.

In the figure of the hunchback Sharon Betcher might have us also recognize the "spirit of disablement" at work—and therefore embrace the embodiment of transient frailties and better failures. This crip theologian and organic farmer has exposed the imbrication of ecology and disablement, of a crip earth and its variously transient bodies. And in that interface she cannot keep her theology hidden in its brilliant darkness: "God here—in this disability theology—becomes ruined, emptied, that nothing-something, so that we are face-to-face with each other—with the sensual flush of sentience and its precarious vulnerability, its injurability."[61] In the brilliant darkness of this ecogod, the ruins of theology lie strewn amidst the ruins of other earthbodies (Glen Mazis).[62] Or, as Betcher develops her theopoetics, "Spirit has been poured into the world to the point that the distinction between Spirit and world has been positively ruined—which is not to say Spirit is either absent or nothing, for 'there is always wind—offshore breezes, scented night-flowering vines—in my cripple' ([Neil] Marcus)."⁶³

Any theology that breathes again, that ripples in the wind and blows us into the public square will, as in the Pentecost story of Acts 2, speak a planetary plurality of tongues, of cultural perspectives and political discourses, oddly meaningful even in their strangeness. Such a pneumatology will have negotiated a secret pact with the newer materialism, only "historical" when that history includes the inhuman times of the universe. In its counter-exceptionalist mattering, a theopoetic exuberance—yes, even sometimes happiness—attends to the precious and precarious flesh of our eternal sentience. God in (the) ruins: consider the peculiar beauty of ancient temples and medieval churches as their stones soften and tumble, as wildflowers grow in their cracks, and no forbidding *potestas* sits enthroned behind the broken-down walls.

Betcher's "nothing-something" alludes to Augustine's rendition, in the *Confessions*, of the *tehom* that opens Genesis. He recognizes there, in a transient moment of hermeneutical pluralism, that the first chapter of Genesis exhibits no simple single creation from nothing but rather supports "multiple true interpretations."⁶⁴ We noted earlier that the deep, when it is not whited-out by the projection of omnipotence, exhibits a beginning without origin: bottomless inception of a fluency without exception. The *ruach* that blows/vibrates/breathes over the face of the waters may arrive in offshore breezes or in climate-heated storms. It was never not material spirit, but the matter of its spirit, the subtlest energy of things, was, if not nothing, never some thing. As it breaks into a chaosmos of becoming creatures, does this spirit of a "messianic nature" positively ruin the Almighty?

In the beginning and in the end, in the messianic new beginning in the face of downfall, God positively fails the standards of sovereign rule that Christendom had erected. As such positive failure comes to be tuned by negative theology, it lets Paul's "weakness of God" register as the plasticity of a precarious relatedness. Forfeiting the projection of causal dominance, the interface of God and world may appear anywhere. Which is to say, everywhere. So, for example, Whitehead's "Eros of the universe" stirs in every "sensual flush of sentience." And, shadowing every downfall, no last judgment but the "judgment of a tenderness which loses nothing that can be saved," a kind of memorial recycling.[65] One imagines then an interplay of passion and compassion rather than command and causation.[66] Rather than the causal once-for-all of the exceptional creation, incarnation, and salvation, we hear, beneath audition, something calling all—always.

GOD OF THE COUNTEREXCEPTION

Alternatives to the lordly omnipotence were repressed all along by the politics of His spokesmen. Any theology that contemplated out loud the counterexceptional all-in-allness could be stomped out as heresy, as pantheism, as revolting democracy. Never mind that the very trope of God *pan en pasin*, all in all, is a Pauline formulation (1 Cor. 15:28). A certain alignment of orthodoxy and power impeded the wide cultivation of amorous alternatives to the Lord of power and might. Passion and compassion ill serve the ends of Christian monarchy or its modern secularizations. They would be the targets of Schmittian mockery—of "eternal conversation," antipatriarchy, democratic chaos, mystical anarchy, revolutionary agonism. Instead of the edgy indeterminacy of the interplay of creatures, the hierarchy of Father, Son, and sovereign earthly representatives kept the order, more or less. God's edge was kept sharp as the sword-tongue of the apocalypse. To rule "in the exception."

So an explicit repudiation of the divine exception is hard to come by. It is no coincidence that the breakthrough of an explicitly counterexceptionalist theology, one taking on the Christian figuration, indeed confronting its politics, happened at a philosophical margin rather than a doctrinal center. Whitehead, however, was unaware of "political theology" as such. Like his contemporary Schmitt, Whitehead goes *theological* just at the moment his gaze sweeps back over the Christian *political* lineage. But with an opposite evaluation: "When the Western world accepted Christianity, Caesar conquered; and the received text of Western theology was edited by his lawyers."[67] We find ourselves here in the final chapter of *Process and Reality*. "The brief Galilean vision of humility flickered throughout the ages, uncertainly."[68]

In its political success, Christendom failed its inner Galilee. Yet sometimes it failed better. It could not erase the imaginary of messianic humility, even if that phrase rings a bit oxymoronic. That vision channeled the anti-imperial potential of the prophetic messianism, funneled through what Bloch called the "Christian social utopias." It continued to flicker, flaming up in all the western movements toward justice, rights, revolution. We noted how Bloch in his Marxism captures the "harzardous business" of hope in the convulsive historical secularization of the Christian resistance to (Christian) sovereignty.

Whitehead zooms in on the formative, idol-bashing period of Christian empire. He captures the betrayal at the heart of the political theology of the west: "But the deeper idolatry, of the fashioning of God in the image of the Egyptian, Persian, and Roman rulers, was retained. The Church gave unto God the attributes which belonged exclusively to Caesar."[69] It is then on the next page of this climactic chapter that Whitehead offers what may be the first explicitly counterexceptionalist theologoumenon: "God is not to be treated as an exception to all metaphysical principles, invoked to save their collapse. He is their chief exemplification."[70]

They continue to collapse. With or without process theology. The Whiteheadian cosmology metaphysically undoes the metaphysics of substance. That is the schema of essentially separate substances, mental or physical, interacting externally to each other, asserting or yielding to mastery. It would harden into the common sense of the modern West, with its propertied men of substance, which Derrida would call the metaphysics of presence. And always behind it, hidden or revealed, operates its theological background. There God's service is required as the exceptional maker of all things from nothing, the white God overcoming black emptiness. Controlling the whole machine from above, intervening miraculously or punitively at will, He lends life, movement, and accountability to systems otherwise trapped in the binary stasis of subject and object. This separate and self-present God serves as holy stopgap, preventing the collapse of the creation schema upon itself.

What Whitehead's deity *exemplifies* is the creative process in which anything actual exists only as an interdependent activity of becoming. No one can take itself out—and God is no exception. Each takes part in shaping the above "shaped togetherness." Each actuality, divine included, is a process of embodiment contracting in itself the entangled materializations of the universe. The metaphor is of divinity as one who does not control any creature, even the electron.[71] An edge of ontological indeterminacy remains irreducible at every scale. Divine agency then does not control the outcome of any becoming, even its own. It *causes* only by calling. It does not coerce, command, or hack its way into the creature. It lures from inside a within in which we all, *pan*, always altogether, "live, and move, and have our being" (Acts 17:28). In this picture, divinity gives itself as the gift of possibility (primordially) and as the reception of its materialization (consequently). It does not manipulate the creatures toward its own ends. Feeling the possibility or screening it out, we struggle, evolve, or collapse. It is not that we are left on our own. We are all in it together, for better or worse: *pan en theos*.

In other words, and usually without God-words, Whitehead unfurls a cosmology of rhythmically intertwined becomings. Provoked at the inception of quantum physics to rethink the whole schema of the world, he depicts a creation that is structured (without exception) in distinct moments of recapitulation: a vibratory universe of microcosmic contractions. "Each atom is a system of all things."[72] All creatures, at every moment of becoming, summarize in themselves their past world from their singular space-time perspective, their now. Their relations are not external to who they are becoming. Each relation to an other is "constitutive." And, in the contractions, contrasts of the past and the possible may intensify, may become complex, may repeat their patterns in emergent schemes of self-organization. Always there is affect, there is life, even at the least complex, so-called inorganic, level.

As the self-organization contracts in increased contrast, consciousness may arise, intentional publics may form, humans eventually among them. The pulsations collect vast differences, and, if these are to be sustained rather than collapsed into mutual exclusion, new social patterns of enhanced complexity arise. At work here we may spot the lure of an amorous agonism, resistant to the antievolutionary habit of binary antagonism. It calls, as the "initial aim" of any moment, for the incorporation, rather than the whiteout, of alterity. (Into only. Hence another.) Then the constituent relations are felt as the potential of the past—for actualization, otherwise, now. As the potential for, we add, creative contraction rather than defensive simplification or fearful collapse in the crisis of difference. (Even gut-wrenching difference. As a homey example: some pale ones of us might take our Republican relatives' views in, no more than we can bear, feel their feelings with understanding, and go forth with a less reactively brittle and more persuasively radical politics.)

Whitehead framed the actual becomings at every level of complexity, electron or Elohim, as cosmologically—which is to say, *nonexceptionally*—partaking of this process whereby the widest and wildest differences vibrate into play. The process schema rules out the notion that difference is best supported by a Sovereign Decider providentially choosing on His own, then imitated by other sovereign and separate subjects. Hence the atheist Deleuze, ferocious philosopher of multiplicity, celebrates the Whiteheadian "chaosmos" of difference repeating and rhizomatically proliferating: "Even God," Deleuze comments, in contrasting Whitehead to Leibniz, "desists from being a Being who compares worlds and chooses the richest compossible. He becomes Process, a process that at once affirms incompossibilities and passes through them."[73] Though technically speaking, Whitehead does not use the language of incompossibles but of multiple possibilities, the point is that the process seeks compatibility in contrast where there had been mere contradiction. It is a process of enfolding a world and unfolding it otherwise. So the process cosmology supports the current quantum meditation we found in Karen Barad, exposing in the smallest bits of matter "an entire tangled multitude."

In the earth, the ingathering has acquired a stunning density. And in that contraction takes place a proto-political gathering. Alluding to William James, Whitehead writes that "we find ourselves in a buzzing world, amid a democracy of fellow creatures."[74] So the entire material world has a political buzz. But he means this in a cosmopolitically radical sense of democracy. It is not because the Creator has *made* it so, creating it in that one primal gesture of exception, from the top down, as a *subsequently* egalitarian universe (to which He would remain the eternal exception). In this vein, Whitehead delivers the following thought, impossible for orthodoxy: God is "the primordial creature."⁷⁵ The process cosmology violates the fundamental dualism of Creator/creature. "The transcendence of God is not peculiar to him. Every actual entity, in virtue of its novelty, trancends its universe, God included."⁷⁶ Our—all of our—mutual transcendence folds into our mutual immanence and out again, differently. The transcending signifies repetition with a difference that makes whatever difference is made. ("To become is the same as to make" [Eckhart].)⁷⁷ "The universe is thus a creative advance into novelty."⁷⁸ Sine qua non of those other creatures, unfathomably and metaphysically *different* from all others, yet this God is not even as First Cause or Creator *excepted* from the creative process itself. This God *becomes*. Not at a particular point of origin, not from nothing, but always, and from within the flux of all becomings. *Pan en pasin.*

This abyssal flux of creativity, Whitehead's "principle of the ultimate," has been meaningfully likened to Eckhart's Godhead, the unground or abyss—signal of an apophatic dark-out of knowability.⁷⁹ Admittedly, most other process theologians do not expressly tune us to this apophatic darkness (flux, depth, cloud, or impossibility) that surrounds any God-talk.⁸⁰ The counterexceptionalism of the Whiteheadian schema, as it breaks up the doctrinal presumptions of Christian theism, lends itself to the experiments in saying that an unsaying permits: deconstructive in construction, metaphoric in metaphysics. If such metaphysics begins in "metaphors mutely appealing," metonyms of divinity included, the appeal of that silence carries ancient mystical resonances. Those are amplified in the Sinaitic sonic system of the "dark cloud" where no cliché of "God" obstructs a widening field of resonance.⁸¹ It might then have reverb with the growing public self-identified as "spiritual not religious" or "nones." It would address them in a language that some of them will be interested to learn has the religious warrant of the apophatic mystics. It is precisely the silent shadow of this discourse that enables "planetary solidarity" with a public vastly exceeding any sort of theism.[82]

That disciplined muteness must find ever more skillful practices of communicative translation. It must plan to face a growing ignorance among secular publics of any but the vulgar and violent versions of religion. At the same time, and with that same recourse to mystery, the constitutive involvement of process thought in natural science helps it keep faith, in contexts as divergent as California and China, with secular struggles for planetary sanity.[83]

Such apophatic plasticity of language also energizes certain margins of a failing theological mainstream, whose churches and temples and mosques may prove to have been crucial to any robust civil society. As the purveyor of the most developed theistic alternative to the controlling deity, the movement of process theology has long fomented local and planetary resistance to the oppressive secularizations, imperial, neoliberal, or dictatorial, of sovereign omnipotence. It does so precisely as a theology that *self-secularizes*. According to Cobb, the founders of all the world religions pushed against the cultic supernaturalisms, indeed the "religion," of their own communities. They taught instead the secularity, by Cobb strictly distinguished from secular*ism*, of prophetic or wisdom practice. As their movements became "religions," the focus on just and mindful material practices often retreated to monastic or radical communities at the social margins. So a political theology of the earth welcomes unconditionally all who join in the practices of earth mattering: all those pre- and postsecularisms that assemble in the solidarity of a worldly *saeculum* of Ecocene collectives.

ALL IN

It is not then that the difference between "God" and "world" is erased. The democratic immanence does not reduce but instead limitrophically magnifies mystery. The deeper and denser the immanence, the more it crowds and clouds into apophatic excess. ("Not knowing how know only no out of.") But then the failures of our knowing motivate ever more learning. We might fail better because we know we do not already know and therefore can learn. If then the divine excess in its negativity does signify a nonseparative transcendence, it is that of the climb-across—*transcendere*—not only within what is knowable in any contracted perspective, but the beyond within, "transimmanence" (Jean-Luc Nancy). The excess of the within marks the novum of the *in*ception, Transcendence then ceases to mark the ontological exception: "It is as true to say that God transcends the world as that the world transcends God."[84]

If all creatures nest materially in the world and metaphorically within the dark difference we nickname God, such a divinity doesn't take itself out but takes them *all in*. (Chris Hayes' *All In* hints at its politics!) Such an All-In can be considered Pauline and panentheistic. It is convenient to oppose panentheism to orthodoxy, on the one hand, and pantheism, on the other (as Mary-Jane Rubenstein well warns in *Pantheologies*.)[85] But in such opposition we not only forfeit the agonistic indeterminacy of an *apophatic panentheism*.[86] We risk at once replaying the sovereign game of us vs. the heretics, but hardening the conceptual line between God and world. This tightening of difference into division becomes incoherent if that God is infinite—without a "fin," a limit that marks an outside; and if therefore the world is within, and ipso facto part of, God. Not separation, only distinction is warranted. The *en* of panentheism, the "within," is what takes the place of the dividing line between creator and creation. To soften, to obnubilate, the line is to discern the cloudy turbulence of all that mediates and distorts a God/world relation.

Thus it was the apophatic tradition of the dark cloud that produced the distinction between the contracted infinity of the world and the negative infinity of the divine, in whom the world is contracted. Such an inference from cosmic contraction as we found half a millennium ago also, and explicitly, beclouds—does not erase—the God/world distinction. Cusa's stunning theocosm unfolds its logic in this way: "Since the universe is contracted in each actually existing thing, it is obvious that God, who is in the universe, is in each thing and each actually existing thing is immediately in God, as it is in the universe. Therefore, to say that 'each thing is in each thing' is not other than to say that 'through all things God is in all things' and that 'through all things all are in God.'"[87] And so *De Docta Ignorantia* describes the infinite as unfolding "in and as" the multiplicity. Once God is considered to be a name for the boundless, then no sovereign boundary between it and the unfolding manifold of creatures will hold. The theologian Johannes Wenck rose to the occasion, publishing an entire book on Cusa's heresy. He argues, in a formulation that nicely captures the orthodox view of what would only later be called pantheism, that Cusa "deifies and annihilates everything, and presents the annihilation as deification."[88] Pantheism and therefore nihilism: this remains the presumption of orthodoxy. To discern divinity in and as all is tantamount to atheism. If God is not the exceptional thing, God is nothing.

Certainly, Cusa was redistributing the notion of *theosis*, "deification," beyond not only its incarnational but also its human exceptionalism. As noted earlier, he describes the cosmos as a whole, the universanimal, unfolding from God, and so as what most resembles God. But the fear persists that such boundless divine immanence, which means a saturation of the cosmos with divinity, entails the erasure of the sovereign difference—and also of the human and Christian political privilege of the orthodox "we." Fortunately, the pope was too dependent upon Cusa to give Wenck much airtime. The Cusan Bruno in the next century was not so lucky. Cusa can rightly be accused of panentheism in a cosmologically radicalized version that undoes the metaphysics of separable substances. But he failed to transform western theology—or, rather, it failed to shift with him. We might read such theological failure as the mark of the way not taken, the way of what might have been. It at once pressures and supports the tensed present. That Cusa's theology did not lead in any continuous sense to process theology makes the contraction of both in an apophatic panentheism all the more cosmopolitically suggestive. Panentheism signifies one thread looping together vastly disparate schemes under a name the originative authors do not use. That would also be true of a more distant ancestor, Paul.

As reported in Luke-Acts, Paul enunciates, at least when speaking to philosophers in Athens, a dynamic panentheism of our living and moving all-in-God.[89] But does it reverse to God-in-all? In Cusa God is "unfolding" in every creature. In Whitehead the divine envisagement participates as an "initial aim" in every creaturely now-time.

If all is in God but God is not in all, the divine functions as a kind of container, in whose hollow the world happens. It could still be conceived as the extrinsic, the excepted. On the other hand, if God *is all in*, shouldn't we be failing a bit *better* already? Even without pantheist unity or omnipotent control? Of course, God being *in all* isn't quite the same as God being *all in*. Paul may be oddly illumining the difference, right back in that first letter to the Corinthians.[90] He is there reflecting on the human interconnectivity: in its "all in Adam," by which a failure went contagious; and then, enliveningly, in its all "in Messiah," as in, "so all will be made alive in Christ," and then, "so that God may be all in all" (1 Cor. 15:28).

May be: God, it seems, "is" not already all in all, in some pure presence. A certain contingency, a deferral, makes and marks the difference. One might read this as an eschatological deferral. Does God at some telos then become metaphysically other than what God will have been until this appointed end? Suddenly immanent to the creation, as before—not? Certainly, the changelessly self-same God of orthodoxy doesn't fit the

Pauline text. But a single, final conversion of God to immanence from transcendence not only (and with our blessing) violates all orthodoxy of His changelessness, it seems also to pose its own arbitrary exception. Perhaps something more theologically subtle was trying to come to voice. "May be all in all." Will possibly be. Not a predictable "will be." Posse ipsum. The "may be" may signify that God can become something God is not always already. But this becoming would happen precisely not as an exception that proves the rule of substance metaphysics. It seems instead to presume a divine becomingness that must somehow always be in play. Thus the anticipated all-in-all of divinity would be the fuller realization of a *possibility* that is always already in all. But then it would not be that the Lord decides or always had decided the time for the End. Rather, we conceive here a divinity who is potentially always "all in all" and, as such, relational through and through, requires our participation. So the all-in-all is only for us experienced as such to the extent that we *realize* it—a "zen-like shift of perception" and so therefore a shift of action, an actualization.[91]

What the metaphor of "sin" always marks is how in our failing of each other we also fail God. And, at least from our little earth perspective, the failing of God by us is indeed a double genetive, a double edge. If we fail to effect a more common good, God cannot do it *for* us. God thus fails to do it. The divine desire for this moment depends upon our limited actualization. By contrast, if we perpetuate a theology of omnipotent control, we can blame all of our failings on a God who might have intervened, might have rescued us. That failure may feel like a betrayal. But the betrayal, it seems after all, is our own. And our theology, our God-construct, is complicit. The failings of the aims of love and justice we call divine are failures of human realization.

Sin thus marks our alienation—hugely systemic, collective, "original," carried in the polis of our self-organization—from what calls us beyond from within, each and all. Invitation or imperative, that calling requires no naming or knowing of "God" to stir some fresh realization, some pulse of conscience or conviction. Our relations to each other can, all in all, foster a more becoming togetherness. Or trump it. So the much remarked unfairness of the notion of "original sin"—of our all having already "sinned in Adam"—comes from its attempt to name the shadow side of our collectivity, our constitutive sociality: Because we are entangled in each other, because of all that has already formed us, our freedom, our ability to realize the possible comes from the beginning constrained. Fear, illness, racism, sexism, even political party are injected before we could in any sense be said to choose. We do not begin as blank slate, ex nihilo. Nor can we extricate ourselves from the vast ecosociality, full of failures, that we depend upon: Augustine's *massa perdita*, a "lump of perdition."

In our chronic inability—blameworthy or hapless—we find ourselves nonetheless—responsible. But in that ability to respond the lump of our shaped togetherness is gifted with much more than perdition. We can participate in its evolutions or revolutions. Grace of the moment, we might each and all make the difference. Just not all the difference.

Theologically, then, our failing of each other means our failing of the All, and thus of the One who would be, who may be, All in All. So grace, like sin, is not a matter of exceptional performance. It is in fresh exemplifications of the All—some gathering and contracting of our collectivity—that we realize, naming it or not, the theos of *panentheos*. "Because all is in all and each in each, God is in all and all is in God." Named or unnamed as "God." And by whatever name, it does not coerce its way in. To put it with feminist bluntness: We do not embrace a God who forces Himself on us. What Whitehead called the Eros of the Universe works uncontrollingly. It lures, and the materialization is up to us, in the failing and shifting togetherness of all our entangled becomings. Paul's All in All may name then not the end but the limit case, the kairotic possibility. God becoming all in all: It begins—on principle, *in principium*, in the beginning that is the inception of every becoming, the lure to genesis. Genesis happens at every scale. The kairos, great or small, takes place here on the eschatos, the verge, of the edgy now.

OH, CHRIST

In its apocalyptic urgency and its militant peace, that eschaton presses upon our earth-time. It squeezes hard against the Anthropocene vulgarity of our sovereign species. We have taken a few breaths with Paul, who does not want us to be anxious. The time shortening in contraction, ingathering, already performs a certain all in all: not in the exception of closure but in the messianic disclosure. It opens the past and the future asymmetrically. If Giorgio Agamben has for this meditation sprung open the connection of the kairos to the *Jetztzeit*, he deftly couples the contraction of time in Corinthians with the figure of "recapitulation" in Ephesians: "to gather up all things in [*en christou*], things in heaven and on earth" (Eph. 1:10). This recapitulation is to happen "for the fullness of time." Agamben finds here a kairos in which "the past (the complete) rediscovers actuality and becomes unfulfilled, and the present (the incomplete) acquires a kind of fulfillment."[92] The unfulfilled history, as in Benjamin's backward gaze, presses the present. And in the Pauline abbreviation this messianic recap contracts with a novel possibility—not a completion but an opening, possibly an outburst. So Agamben makes audible the resonance between the messianic recapitulation and Benjamin's "monstrous abbreviation."

If all, *pan*, is always already *en theou*, what happens *en christou* figures its inception—not as exception but, to the contrary, as fulfillment, that is, as realization. This means at once to actualize a possibility and to reveal it. Agamben's reading thus helps to articulate a christology that does not supersede, that does not except itself from the All, but that takes it all in, contracts it all in itself, in and as the messianic event. It is for this climactic recapitulation that Paul uses the language of the kairos, not chronos: *pleroma ton kairon*. So, in Agamben's rendition: "Having just laid out the divine project of messianic redemption (*apolytrosis*), Paul writes, 'as for the economy of the pleroma of times, all things are recapitulated in him.'"[93]

I have elsewhere offered an apophatic reading of a gloss of this Pauline "work of recapitulation, summing up all things." God, for the second-century theologian Irenaeus, "in an invisible manner contains all things created," and so "the Word of God . . . came to his own in a visible manner, and was made flesh, and hung upon the tree, that he might sum up all things in himself."[94] This figuration carries the glimpse of redemption not as supernatural takeout but as immense intake. As feedback loop of the past in its now-time, it suggests a cosmopolitical expansion of Foucault's "history of the present." It is one great recycling of the world, known and unknown, of all its temporary realizations, its endless failures and waste. Indeed, Agamben finds Paul's short verse on the *oikonomia* of the pleroma of the kairos so dense with meaning "that one could say that several fundamental texts in Western culture—such as the doctrine of apocastasis in Origen and Leibniz; repetition or retrieval [*Gjentagesle*] in Kierkegaard; the eternal return in Nietzsche; and repetition [*Wiederholung*] in Heidegger—are the consequences of an explosion of the meaning harbored within."[95] I recap this recapitulation of readings of the *recapitulatio* not only with the ancient exegesis by Irenaeus added but the *complicans* of Cusa and, as influenced by Leibniz's microcosmic monad, Whitehead's actual occasion as recapitulation of the past universe, as well as Deleuze's sense of repetition and its folds as it iterates Leibniz, Kierkegaard, and Whitehead together.

Whitehead did not resort to a language of Messiah or of Christ. In his cosmology, the most natural, the cosmological, rhythm that brooks no exceptions iterates the history of the universe from each new space-time perspective, each becoming creature. But it does all replay itself in the "consequent nature of God," which, moment by moment, enfolds all things in itself as the divine becoming. And it issues in the next moment as divine lure, incipient in each occasion, however much or little *fulfilled*. As Cobb translated the divine lure into christology, Christ as logos names—far from the sovereign Son of the omnipotent Father—the call to "creative transformation."[96] In Cobb's lineage of process theology, resistance of the U.S.-dominated global politico-economic order for the sake of a common earthly good counts as messianic work of creative transformation. It resists all subjugation of creativity to the image, the doctrine, or the name of Jesus Christ. But this struggle takes christological shape, named or unnamed, depending upon context. The messianic oikonomia challenges the Goliath of global economism.

When Agamben makes audible the resonance between the messianic recapitulation and Benjamin's great abbreviation, we hear a related material economy at play, cosmopolitically, intensified. It is under the pressure of that summation, or crystallization, that the door might crack open—at any moment—to the messianic event. And the christological rendering of that pleroma, trapped for a two-millennium history in a mistimed closure, a tragic miscarriage mistaken for a triumph, points in this rereading to the way hardly taken. Its truth and its life are dissipating and now threaten to burn out.

If Christendom—the sovereign Christianity—has failed Christ, does that not mean also that Christ failed us? Surely, as has "His Father," inasmuch as we surrender to the regnant christology.[97] If, however, one attempts to pick up that way again, to recycle the ancient sources through a more honest hermeneutic, it is not as though the way is clear. One quickly stumbles upon a language of enmity. For instance, after the passage I had warmly cited, Irenaeus continues thus: "[Christ] has therefore, in His work of recapitulation, summed up all things, both waging war against our enemy, and crushing him who had at the beginning led us away captives in Adam. . . . As our species went down to death through a vanquished man, so we may ascend to life again through a victorious one."

One can decode the symbolism in the context of an endangered second-century minority, militant in its courage. Indeed, that movement was unflinching in its critique not just of its culture's but of our species' surrender to "the Enemy"—the empire, evil, Satan, sin in its systemic force, its ruling world schema. But there is no escaping the antagonistic binarism of such rhetoric. The Schmittian friend/foe apparatus is working its way up through the earliest theology. As these symbols so evidently still live and work within the political theology of sovereign aggression, one may credibly go—and so many ethically do—the dump-Christ route.

If however a political theology of the earth would remain committed to the less wasteful hermeneutics of recycling, how would we shift this Christic antagonism at its source? How might our respectful agonism liberate the recapitulatory christos from captivity to Christian triumphalism?

Not easily. Indeed, a flagrantly theopolitical violence bursts out already in the *Urtext* of the recapitulatio, 1 Cor. 15. "When he hands over the kingdom to God the Father, after he has destroyed every ruler and every authority and power. For he must reign until he has put all his enemies under his feet . . ." Besides spooking feminist theologians, such language no doubt provided sanctity to every Christian we/they power play. Yet it is telling how even such a Pauline theologian as N. T. Wright, hardly identified as progressive, nails the politics at play in that passage: "The return of Jesus from heaven to earth, the parousia, was formulated, probably by Paul himself as the earliest

Christian thinker known to us, in conscious opposition to the parousia of Caesar." We had earlier noted the subversive interplay of the "kingdom of God" with the sovereignty of Caesar. "The christological titles Paul uses here for Jesus (saviour, lord, Messiah) are blatantly counter-imperial, with the word 'saviour' in particular, used here for the only time in the normally accepted Pauline letters, echoing around the Mediterranean world with the claims of Caesar." Historical scholarship makes clear the importance of recognizing the pre-Christian political theology of Caesar, who called himself savior, indeed "divine son of God, redeemer, lord . . ." Wright captures the ironic edge of Paul's rhetoric: "God's plan for the world is thus, in Paul's mind, the reality of which Caesar's dream of world domination is the parody."[98] Christian imperializations of sovereignty—"the deeper idolatry"—would soon amp up the parody. And recently we have endured an American burlesque of its white supremacist fire and fury. Might we who still have truck with the anti-imperial messiah fail *better* than we have so far at circulating, sometimes secularizing, a counterexceptionalist christos?

In the troublingly lordly passage presented earlier, Paul is actually exegeting an ancient psalm, trying to locate his messiah in its message: "But when it says, 'All things are put in subjection,' it is plain that this does not include the one who put all things in subjection under him" (Ps. 8). Is this just Paul trying to inject into Hebrew monotheism a new Christ-exceptionalism? If so, it is all the more surprising that Paul immediately pulls the rug out from under its sovereign feet: "When all things are subject to him, then the Son himself will also be subjected to the one who put all things in subjection under him.

In other words, in the eschatological pleroma this Son is no exception; *he is also subject*. Corinthians 15 thus served as the proof text for what was called subordinationism (kin to the Arian heresy but with a longer reach). The text could be read as supporting the sovereignty of God the Father over the Son, even with

Christ taken as divine. We may for now spare ourselves the old patristic debate on subordinationism and the trinity. But it is not without political significance for contemporary theology. Jürgen Moltmann demonstrates the link of subordinationism to the "monotheistic monarchicalism" with its "uncommonly seductive religious-political ideology." He finds here "the fundamental notion behind the universal and uniform religion: One God-one Logos-one humanity; and in the Roman empire it was bound to seem a persuasive solution for many problems of a multi-national and multi-religious society."[99] We noted Bishop Eusebius' enthusiastic support for the monarchy of Caesar combined with the *monarchos* of the Father. We might want to avoid identifying monotheism, so readily identified with Judaism and Islam, as the oppressive guarantor of a political theology of imperial unity. But Moltmann's target is the long history of *Christian* monarchicalism, with its effectual sabotage of the relational egalitarianism he finds inherent in the trinity.

More recently, Linn Tonstad writes the trinitarian persons as "relations oriented to each other in the vivid life of God."[100] She rejects not only subordinationism but, refreshingly, any fixation on the "trinitarian persons and their constitution." The borderlines between the triune persons become, we might say, limitrophic relations. The original African logic of the trinity as "different not divided, distinct not separate" (Tertullian) has a fresh chance.[101] Such intensification of the inherent relationality of the trinity liberates the "personae" from the exceptionalist incoherence of the three as one (simple) metaphysical substance. In her "queer theology," Father, Son, and Holy Ghost do not vanish, but flutter loose from the iron triangle of their masculine sovereignty, free at last to join a multitude of planetary metaphors recalling us to our divine entanglement. Which calls us to embody it anew. Building on the noncompetitive creator/creature relation of Tanner's politics of God, Tonstad lifts up "the unrestricted community that God establishes among humans when God establishes communion with humans."[102]

The implications for political theology in our present wall-building moment are manifest; Tonstad invites the nutritive excess of "Banquets without borders."

Paul, however, had no trinity as such; he is struggling with his own monotheism and against a subordinationist hierarchy. Those put "under His feet," in the cited psalm, are those who do the forcing, who run on enmity—the whole schema of ruling sovereignties. What if we read "when all things are subject to [God] then the Son himself will also be subjected" in terms of the verse's figure of recapitulation? Then the adjective "subject" cannot in Paul's exegesis mean "is now inferior to or dominated by," as in, under the feet of a ruler. Rather, "subject to" in the Pauline hermeneutic would signify becoming enfolded in the one who enfolds—recapitulates—all things.

"So that" at last, in verse 28, "God may be all in all." That relational plenum is what all are being "called"—not forced—into. We could argue in other words that despite the regrettable ambiguity of "subjection" and the antiquity of the irony, it is the messianic counterexception that unveils this all-in-all as a cosmological, and so radically ecosocial, inception.

AMOROUS AWAKENING

An earthen political theology requires not just an ethos but a cosmos of interconnectivity. Any creature holographically embodies its world, with vastly varying degrees of detail and relevance, of proximity and awareness. So the recapitulatio would not curate a metaphysical anomaly but a novel realization of something always and everywhere in play. But here the messianic body contracts its members, their worlds, in amorous fullness. Not in order to take believers out of body, out of world, but to take us *all in* deeper.

At that depth, in those creative waters, new experiments might renew our cohabitation of the Ecocene—economically, ecumenically, ecologically. Jesus' particularity would *exemplify* the mutual participation that always already constitutes us as creatures, but that in the messianic event was realized not in a metaphysical exception but in a peculiar fullness, in a body that bespoke an endless all-in-all yet to come. More old christological debates rear their head: Anselm's substitutionary atonement versus Abelard's christology of the moral exemplar. The latter was strongly rejected by the Reformers as mere works of righteousness. For Christianity had by early modernity gotten trapped in the false alternative of humanist works of self-salvation versus reliance upon the grace of omnipotence. But the way of the exemplar, the groundless middle of a graceful synergy, did not disappear.

The messianic exemplar may perhaps signal the human coordinate in the moral arc of the universe. If the human may with everything else be "summed up" in Jesus' mattering life, it is through its antihierarchical connectivity: "What you do to the least, you do to me" (Mt. 25:31–46). Keep the "Lord, Lord"—*I* am not the point.[103] No sovereign subject survives this radical entanglement. Such amorous agonism on behalf of the vulnerable brooks no reduction to moralism, purism, identitarianism. Rather, it offers a biblical amplification of an intersectional ethics for a time of unveiled white supremacism, Islamophobia, climate denialism.

Does such cooperative gospel grace draw already upon the pleromic all in all? Getting over my Jesus versus Paul habit, I am guessing that, under the pressure of the now-time, no ethic can hold, can breathe, without the amorous spirit of that fullness. And, as Taubes insists, surprisingly: far from diluting Jesus' rendition of "love the neighbor as yourself," Paul radicalizes it. He makes it the whole summation of the law. "For the entire law is fulfilled in obeying this one commandment, love the neighbor as yourself" (Gal. 5:14). Taubes' point is that Paul is deliberately, "polemically," not adding the "and God" of the dual commandment.[104]

L. L. Welborn nuances this reduction of the love commandment to the "neighbor," which becomes "the other" (Rom. 13:8b), as the "possibility of an 'awakening.'" The awakening would occur when believers, roused from the "sleep" imposed by the imperial sovereignty of the dominant order (schema), grasp the full import of the messianic event for their present life. In its eschatological radicalization of the community, its "constellation is the 'now-time' that believers must discern."[105]

The kairotic possibility of "radical hospitality" means to "welcome one another, as the Messiah welcomed you" (Rom. 15:7). It climbs across the differences of race, class, or gender: "There is no longer Jew or Greek, there is no longer slave or free, there is no longer male and female; for all of you are one in Messiah Jeshua" (3:28). This collapse of hierarchial identities has been admired by such a firm atheist as Badiou as Paul's "militancy."[106] This means for him the politically utopian and revolutionary call to "consolidate what is universal in identities." With a more democratically radical Paulinism, Welborn checks Badiou's antipluralism. "Thus the 'communitarian particularisms' against which Badiou rails as impediments to a universal truth, are not erased in the Messiah, but embraced and affirmed. 'The one who loves the other,' the other as the very embodiment of difference, is the one who 'fulfills the law' (Rom. 13:8b)."[107] Coalitional intersectionality thus finds flesh and fullness in this awakening.

We find ourselves entangled in one another no matter what. But in amorous agonism membership *in each other* becomes a calling, a responsibility, a new creation. So love in its radicality does not then work as an exception to the law but for its "fulfillment." This is a crucial point for a political theology, as it tunes its secularizations to their messianic edge. The specifications, formulae, and rules of the law may be suspended if they come into contradiction with this "one commandment"—but that is not to move to a sovereign state of exception, above the law. To fulfill the law does not mean to stand above it or to supersede it (or its later racialized people). It means to realize the *spirit* of the living law, as the ruach that always blows through and beyond the letter.

Thus awakened, law supports new self-organizations of a public in the face of critical difference, a public facing crises of difference. With apophatic fluency, the love commandment then translates into queerly diverse discourses of public responsibility, social justice, and sustaining ecology. It eludes sentimentality and goes systemic, holding politics to the better angels of its history. Its spirit flows between and beyond political systems even in their downfall, brooding over precarious nests of possibility. Whatever its moment, its public comes out of Christendom's privacy trap, in which love sticks to its own. Those neighbors become whole populations, human and not human.

Stripped of certitude, unwhiting-out its spirit, who knows, the Christ-event may outlive its platitudes. Unknowing better now: in its mindful hiddenness—first of all in the bodies of "the least"—it feeds fragile and insistent new alliances. It does not seek conversions to any Lord. In its messianic inception it deconstructs its own paling history of sovereign success. It exposes the cruel optimisms of Man's triumph, Christian, nationalist, or capitalist: worstward ho.

The labor of amorous agonism exemplifies the noncompetititve assemblage of a planetary public heading into trumped futures. Meaning to fail better, its theology does not except itself from its own context: Christian now only as ever more mindfully entangled in Jewish, Muslim, Hindu, Buddhist, and indigenous wisdom and imbricated in the secularreligious translations that will let us coalesce. Us the new public of this ancient earth. All in all.

We have here undergone an experiment in a kind of theology that bends, hunchbacked, toward the earth. It means to be practiced in the press and crunch of time. Such a political theology nonetheless slows time into contemplation. The moment is pulled back by lament, tugged ahead by hope, and widened in the pangs of its contractions. Losses past and losses to come multiply; but so do the bodies who *mind* them and refuse their inevitability. The assembling of an interlinked public, its participants "members one of another," was always a synergy of mourning and joy, of world-loss and the mobilization of an other world. Not otherworldly. But other than this one.

The amatory spirit does not cease to enliven those who will breathe it in. It blows through any religion, any irreligion. Nonetheless, our charismatic alternatives can get crucified. Their resurrections may fail to materialize beyond their circle of compañeros. Their spirit may howl across deadened landscapes. So long ago, the creativity of our geneses got us in over our little human heads. And now the vortex of our variant apocalypses—this meditation tracked just three—keeps us at our edge. The failing schematisms of politics, of ecology, and of religion do not merge with each other, but they tangle inseparably. Theology, as political and earthen, keeps faith with the entanglement. It may actually offer an opening into what otherwise remains a shutdown.

Even if the love we occasionally nickname God does not fail us, we as a species may in some final sense fail it. We do not know. But this we do: in the inception, which has come, is becoming, and is to come, a better now is possible.

APOPHATIC AFTERWORD

A conclusion performs its own time-tensed contraction. It writes toward futures it does not know. Its now-time glimmers in the dark. In this the present of a text is like that of its widest context: the moment folds into history, its present unfolding possibilities that may only find realization enough to deepen tragedy. Or that very depth may ripple and roar beyond disappointment into unpredictable mobilizations. Certainly the present work has been written without expectation that the contractions will have lost their convulsiveness.

If amidst the death spasms of dominant schematisms of democracy, of planet, and of theology we discern birth pangs, we attend here differently to the contractions. So, in service to that difference, this political theology has tried to read the tensed present of the earth from the perspective of a theology that never wrote toward The End. It lacks on principle the certitude of such hope or such surrender.

The unending therefore may require an exercise in multilateral unknowing. This form of mindfulness, practiced in Christian history as negative, or apophatic, theology, made itself known in the last chapter as entangled in the matter, the materialization, of "God." What will help us to recapitulate the entire project now will be this proposition: in a certain sense, *political theology is already always negative theology*. Secularization is a kind of apophasis, an unsaying. But then it needs its own *docta ignorantia*, its knowing what in its moment it does not know. The political apophasis does not silence all traces of religion but listens for their reverberation.

If we are as an emergent public to fail better, we will practice attention to the heterogeneous affects and effects of religion. Then we mind—with hugely varying degrees of engagement—theologies past and living, better and worse. That will require imaginative secularreligious entanglements at the edgy limits of our coalitions. There the theology unspoken but disclosed in the secular can help us, with Beckett, to unknow better now. Which practically means: at the moment of crisis, we might together take risks that matter. Across the loom of critical difference, we might release the speeding shuttle of *kairos* into actualization.

A DARKER BRILLIANCE

The argument of this book has proceeded from the juxtaposition of the power of the sovereign exception to the potentiality of an ecosocial inception. This contrast registers affectively as the difference between the politics of antagonistic unification and the coalitional capacity of an amorous agonism. Against the friend/foe politics, we practice the political as the self-assembling of an insistent public at the edge of chaos. In the Ecocene, the economic malpractice of our national politics is beginning to take us over the edge and into an earth no longer friendly to the schemes of human civilization. They pitch, in the meantime, new lines of self-destruction. I do not know what floods, fires, and furies, what extinctions or exterminations might have been proceeding in the interim even of this publication. For now, they are adding up not to the world's end but to the end of the *civis*-supportive Holocene—an era that will be appreciated better late than *ever*. Provoked by emergency, the chance may also be growing, against all those odds, of an Ecocene emergence: a time of agonized attention to the interdependence of humans with the planetary plenum of nonhumans, not to mention of humans cast as subhumans. To outgrow our anthropic exceptionalism may, oddly, mean outing the string of white-heterosexist-nationalist-classed-religious exceptionalisms that so unexceptionally rule. Might the multiple entanglement of these crises then twist, might it be twisting even now, into the intersectional force field of the inception?

The language of the foregoing paragraph, you might notice, remains almost entirely imperceptible in its theology. Yet this book has argued that theology—failed theos, fading institutions and all—might help our motley public, in its politics and in its planetarity, to fail better. Because a theology of omnipotence comes already larded in Western politics and largely hidden in its operation, the countertheology may prove indispensable to the resistance. Seen and unseen, it relinquishes the theopolitical habits of exceptional power and supersessionist truth. The unconditional cannot be found in the unquestionable; what unconditionally matters materializes in mindful uncertainty. The messianic possibility that it carries for an ecosocial arising precedes and exceeds any hope of victory, whether Christian, nationalist, or capitalist.

As possibility, it sprouts in the darkness. Its roots tangle with a long spiritual history of unknowing better—better first of all than not knowing that one does not know. This is the apophasis we considered in the last chapter as a logos (no mere silence) of theology itself: the discourse of its own unspeakable edges, the *eschatoi* that escape cognition even as they demand recognition. The unspeakable "name" of one Hebrew nickname for God—*Hashem*, "the name"—captures the irony; the "brilliant darkness" of Gregory of Nyssa surged into the radical "not one or oneness, divinity nor goodness" of Dionysius the Areopagite, where the theopolitics of sovereignty crashes: "It has no power, it is not power. It is not kingship."[1] The many articulations of negative theology take aim not at the image of God but at the idol: the deification of any human idea, image, definition. The aim was never to erase—to whiteout—theological discourse but to let its darkness shine.[2] Politically speaking, and therefore speaking in the unconditional interest of an ecosociality of justice flourishing across the multitude of whited-out and silenced constituencies, those of us oriented to some religious language or another will then neither simply repress nor simply assert it.

We may instead let theology's own postmodern darkness become luminous, let it work for rather than against our illumination; we may let its own silences *signify*. For political theology itself signifies precisely the theology secularized—which is to say camouflaged, unspoken, and operative within modern political concepts. Therefore political theology can be described as a mode of negative theology, whether it *knows* itself as such or not.

Even granting such a proposition, however, one might ask: What is the use of calling the very silence of the secularized theology—whether in notions of conservative sovereignty or of messianic justice—*theological*? Why not, once having recognized it and taken responsibility for its politics, just let the hidden theology stay buried and focus on its worldly effects, both oppressive and liberative?

And I would answer: attention to the unspokenness of the theology—and so to the theology of the unspeakable—has everything to do with the possibility of collective inception. For such an apophatic practice gains access to the oneiric space surrounding and infiltrating language. One tunes here to the unknown not in speculative distance but in the operation of ambiguous affects, aesthetic ripples, and sudden outpourings, of something between a collective and a political unconscious.[3] Here can be negotiated the lasting "moods and motivations" (to return to Clifford Geertz's classic definition of the work of religious symbols) that drive the practice of a collective. The possible contagion of an amorous agonism carries, for instance, a political theology in direct resistance to that of the consolidated antagonism of the religiopolitical right. The way of its public ethic is prepared by that unknowing that *trusts* where it cannot know, *hopes* where it cannot predict, *loves* where it cannot agree. (Pardon my faith-hope-love paulitics.)

On that way, both nihilism and denialism, certainty of extinction or of redemption, secular or supernatural, can—just possibly—yield to the darker brilliance of the now-time. We have considered a triple apocalypse of the political, the earth, and theology. But we discerned instead amidst the indubitable dance of triune doom the possibility of shiftings local and planetary. In that kairos of present possibility, does apocalypse go apophatic? Does it reveil in indeterminacy what it seemed so garishly to reveal in prediction? What if it is through the veil of a "hope draped in black" that it dis/closes what it had seemed to close down?

So then let us in haste run our recapitulation in and through the darkness of that very opening, into a triple practice aftershadowing the three chapters: of a political apophasis as assemblage, as animality, and as action.

APOPHATIC ASSEMBLAGE/ THE POLITICAL

The apophatic emerges from an ancient mystical practice soon marginalized by the orthodoxies of the Abrahamisms and readily signaling affinity to the spiritualities of Asia.[4] Indeed, it provides in the 1453 *Peace of the Faith* arguably the first explicit reach toward religiocultural multiplicity: If by definition no finite one "knows" the infinite, then neither can any faith be excluded.

What is "sought in the different rites by different names" remains "unknown and ineffable to all."[5] Crucially, this interreligious potentiality was articulated not for an abstract pluralism or polite ecumenism but as an argument for planetary peace. At that particular moment, Catholic Europe was threatening a new crusade, fearing invasion by Ottoman Islam, which had just taken down the Byzantine Empire. In other words, negative theology long ago went political.

The apophatic unknowing cultivates hospitality to shared uncertainty. Its practice can host in its very darkness collective indeterminacies, whether unspeakable in horror, in precarity, or in wonder. As a minimal form of apophatic practice, the moment of silent breathing together (*con-spiratio*) already begins to gather a public across difference. A force field of speechlessly breathing bodies requires no "them" to tune its "us." Only sometimes would we name that energy *ruach, pneuma*, breath of the world. The "spiritual not religious," the "nones," the agnostics, even the most militant of atheists are usually glad to catch a breath in shared silence—and so to stretch the moment.

The most practical potentiality of an apophatically darkened theology is for the gathering of a public across critical difference—and so for the political itself. Thus Connolly had made the claim for political theory that a "fungible element of mystery," irreducible to any religious or irreligious beliefs, can broaden the bandwidth of the coalition needed, at least in this country, in the struggle against "aspirational fascism" and for "multifaceted democracy."[6] Might theology in its affective agonism then *mysteriously* intensify the aspiration, the con-spiration, of ecosocial assemblage? And, in that spirit of deeper, darker solidarity, might such an intersectional pluralism intensify its own difficult sociality?[7]

This would mean facing not just the crisis posed by the trumped U.S. right to the whole democratic spectrum. It exposes also the critical difference between liberal, often neoliberal, and *social* democracy, with its historic links to a yet more radical democratic socialism.[8] So, does democratic radicalization sabotage the coalition that must work vigorously within and through electoral politics? That would be the objection posed by those who prioritize (also for the sake of a wider coalition) a "moderate" democratic positionality. But the generational trend shows otherwise. Democratic Party dogmatism will not galvanize the youthful public that supported Bernie Sanders; they would rather drop out than sell out. Nor will party business-as-usual mobilize the critical difference of African Americans or, *very* differently, of the white working class. The broad democratic spectrum is beset by contradictions.

Farcical 45 may have been galvanizing his "we" by force of sheer enmity. But the opposed "we" cannot compete in force, fury, or in funding with the totalizing antagonism of the right. If then we refuse to acquiesce in an answering unification by antagonism, we also refuse to read the toxic charisma of one president as the great exception to U.S. politics. We instead read him as dire symptom of an anthropic race/class/gender/sex condition that has from the start infected U.S. democracy. Does a fresh inception become more likely? Not as a new purism but as a self-implicating complication? Perhaps, too, then, some blue waves of democratic agonism can sublimate the antagonism within our promisingly immoderate left between, in particular, race and economic analyses. That means relinquishing not the rigors of radicality but the purities of orthodoxy.

For instance, no broad enough U.S. public will ever yield to socialist orthodoxy. So I find stirring the incipience of an "apophatic Marxism." It bursts into articulation in a subtle voice of European socialism, the renowned writer China Miéville. He opens the way of "a political via negativa, an apophatic revolutionism."[9] Well-read in the Catholic mystical tradition and no theist, Miéville links this unsaying not to a political theology of revolutionary salvation but to what he calls, also as the name of a journal, *Salvage*: "Edited and written by and for the desolated Left, by and for those committed to radical change, sick of capitalism and its sadisms, and sick too of the Left's bad faith and bullshit."[10] There you have it. This political apophasis will not shut up and shut down. In linking the negativity of Marxist critique to the theological unsaying of Christian bad faith, it dissipates the exceptionalist BS of progressive certitudes. In "the paradox of actually existing revolutions" as against the "potential for utter reconfiguration," Miéville situates within the historical now-moment and "precisely beyond words, a messianic interruption—one that emerges from the quotidian."[11]

In the present context of crisis, no progressive movement can deliver on its messianic promises if it does not first deconstruct its guarantees. Then it might be "unknowing better" its own possible truths. Unbelievable faiths might give way, even by way of desolation, to some trustworthy tangles. Who knows, even to the coalescence of an immense ecosocial manifold, unafraid of the animality of its ecology and the socialism of its democracy. There may be nothing more important to the salvaging of a habitable earth future—beginning in the emergency of now—than the art and practice of apophatic assemblage. Its temporal contraction, sunestalmenos, ingathers a messianized mess, yes, maybe even a mass, of difference. Might we yet counteract the secularized theology of the sovereign exception with the political theology of the emergent inception? At such an edge, "messianism and mysticism are twins."[12]

APOPHATIC ANIMALITY/EARTH

How, though, does political theology in its apophasis turn to earth? It is at least a matter of planting in the political unconscious ecotheologies of creation—old and new spiritual schematisms of divinely mandated earthcare (*laudato si'!*), of ecospirit, of sacred earth—to materialize in secular form.

There is, however, a denser contraction to this earth apophasis. Might its speechless breathing take us deep into our own earthbound bodies? Does its silent spirit-breath not animate the flesh we share with "everything that has the breath [or spirit] of life" (Gen. 1:30)? So we ricochet here back to precisely the passage used to justify the most overbearing anthropic exceptionalism. As we have read them together, the Genesis verses 25–30 present the normally ignored climax of the sixth day: the original (originally vegan!) dietary code proposed—without exception—for all of us breathers. And by habituated sleight of hand that same passage turns into justification for our species-exterminating recklessness in the name of our sovereign "dominion." The anthropic denial of our shared animal breathiness turns into the breathlessness of Anthropocene climate denialism. As I write, nuclear physicists have set the Doomsday Clock forward two minutes to midnight, having factored in Trumpist nuclear recklessness and global warming denialism.[13] But when the nuclear threat recedes again, the climate clock will only tick forward.

That breath of life signifies our interdependence with all that lives in the cycles of oxygen, carbon, nitrogen . . . Here another apophasis comes into play. The myriad other creatures of the earth do not *speak*. They communicate, they signify, they have something like languages; they perform varying degrees of articulation and cognitive sophistication. But if they "speak" it is an apophatic sort of talk! Keeping in mind Massumi's warning against drawing any firm line between us and the other animals (as, for example, at the threshold of language), speech does seem pretty much a human practice.[14] This is not a matter of tightening the primal us-them. To the contrary: by entering into a shared register of unspeaking, might we humanimals unknow better now—and first of all our fellow critters? Increasing volumes of research demonstrate the extent of the apophatic forms of communication between, for instance, plants: as trees have been found "to use a network of soil fungi to communicate their needs and aid neighboring plants."[15] Or, for example, Anna Tsing's matsutake—*The Mushroom at the End of the World*—which, "in the ruins of capitalism," help damaged forests to grow again and displaced people to survive.[16]

A negative theology of the earth might then make possible modes of coalition for the sake of our environmental salvation not just with other humans. We might respire and conspire with the nonhumans in us and around us in order to salvage the precariously interdependent life of the planet. Though such interspecies solidarities have an immense history within indigenous spiritual traditions, they may only happen now through new, experimental, even, as noted with Massumi, playfully animal, interactions. These practices can be called apophatic animalities.

In view of "new givens and new unknowns," we must somehow "inherit the trouble, and reinvent the conditions for multispecies flourishing, not just in a time of ceaseless human wars and genocides, but in a time of human-propelled mass extinctions and multispecies genocides that sweep people and critters into the vortex."[17] The vortex may persist even if we do not desist in the reinvention. Staying with the trouble means letting go both the chilling optimism of any technofix and its shadow opposite, the surrender to all the too-lates. It will require new skills of interspecies translation. We then practice an animating speechlessness that shades into its own unknowing, not just of a divine infinity but of a virtually infinite world of material finitudes. That boundless cosmos presented itself in *de docta ignorantia* as animal. We might listen with a new and silent care to the communications of its manifold members. Which are also us.

Always still that "animal which therefore I am" folds me silently, in the midst of my chatter, back into an earth-full of nonhumans you and I never left. Our animality exposes us not just in sensuous play but in violence and lament to our shared vulnerability. It is the attention to the *vulnus*, wound, that gets concentrated in the figure of the crucifixion. First Testament scholar Stephen Moore attends with exquisite animality to the nonhumanity at stake on the cross.[18] In his reading of the Book of Revelation, Jesus appears repeatedly and consistently as the "lamb bearing the marks of slaughter." Moore concludes that "Jesus' humanity flickers indecisively in Revelation and is ultimately eclipsed by his animality."[19] The messiah is "a butchered sheep: it has met the fate of sheep everywhere."[20] And so, in this stunning ecological rendition, the human who challenged the sovereignty of an empire is put to a politically public death: "an altogether abject death, an utterly dehumanizing death, a death more fitting to an animal than a human."[21] It might overburden Moore's quadripedal hermeneutics—which soon thereafter unfolds as divine/animal/vegetable/mineral—to attribute to it a political theology.[22] But in emphatically deconstructing the sovereign exceptionalism of the imperial Christ, his divinanimal Jesus performs its own almost speechless politics of new creation, right at the troubling climax of Revelation.

So the apocalyptic vortex twists open in apophatic animality. Whether or not we stay with the trouble of christology, let alone any theology of the cross, we will be haunted by its counterimperial risk on behalf of all that breathes.

APOPHATIC ACTION/THEOLOGY

As the verbosity of chapter 3 attests, it is not that theology turning apophatic is turning speechless. It is fair, of course, to suspect that apophatic theology as a present procedure merely wraps the harsh death of God in a postmodern specter of uncertainty—a corpse cloaked in a stylish Goth aura of mystery. I hope that something livelier is coming through. Perhaps it does convey a hint of retromonastic stylization. For negative theology surely carries a deep history of contemplative practice. Its silence shadows any theological assertions but does not erase them. It clouds them with possibility sometimes so dark as to signify: impossible. And then it goes darker still, into the possibility tangled in impossibility, *posse ipsum*, where the blackness shines.

As political theology, a theological assertion becomes secular in the struggle for material actualization. It goes dark as theology precisely in order to act. As *negative* political theology, it transmits its darkness mindfully. Dark flesh shines with new meaning, inseparable from the dark earth and embodying new im/possibility.[23] Whether called theology, then, negative or otherwise, a sense of sacred responsibility, of ultimate accountability, shadows our earthen actualizations. Its negativity is never "just" theological, but darkened in lament, protest, and the critical edge of crisis. And so it attracts our becomings, our emergence together as a new public. Entering the cloud of our own now-time with eyes open, the agonism deepens amorously: We struggle through and beyond the toxic antagonism that shames so much of the religiopolitical past and threatens so much of the planetary future.

So then there appears another meaning of the theological unsaying that has not quite been *said:* The silence signifies not just a practice of attention but of *action*. The theory of theology flows or flames—if it lives—into actualization. It takes place. No speech will take the place of the deed. The deed may require great torrents of speech. But as theos logos, it seeks enfleshment in the common world as the most common good, reverberating through the variegated skins of the undercommons. What is required of us? "Do justice, love mercy, walk humbly with your God." A slice of local practicality, those words were for years my theological school's mantra. Increasingly it was printed "Do justice, love mercy, walk humbly." Not because God had died but because practice takes the place of cliché. Those who want Godtalk know the rest of the sentence, and those who don't are welcomed. Now the tagline reads, "Rooted. Innovative. Courageous." At the moment of writing, our alum the Rev. Dr. William Barber II was on his way to train the community in the political activism of "moral revolution."[24] The practice continues.

So the kind humility of Micah's walk does not resemble passivity or subservience; we feel also its impatience, as in the gospel parables, when the fig tree, the seed, or the steward do not bear their fruit. Theology can pick texts and reseed them, can plant and nurture. But it cannot do our fruit-bearing for us.

No more can its theos, who needs our actualizations. In order not just to *be* but to *act*. No longer the magician of the miraculous intervention, God sounds—silent. We may feel abandoned. Therefore, we have considered the counterexceptional theology of a divine entanglement in the becoming of each and all. Becoming all in all, it suffers and enjoys us each (yes, you are a marvel)—but not in abstraction from our all-togetherness. We are each contracted in that All. And yet each creature contracts all to its own perspective, "the universe in a blade of grass" (Whitman).[25]

That all-in-all does not dictate the singular shape and limits of any assemblage. Of any blade of grass. Far from cosmic dictator, the "word of God"—*logos theou*, theology—communicates in the grassy silence. If it may be said to translate on occasion the unknowable into our unknowing better, it is to call, to provoke, to invite a fuller realization. Ancient of days, hope of better days, lure to the loving justice of a new city, heavens and earth, a new day, it calls us out of our protective privacy. Contracted in our depths, it calls us to a becoming public. A public becoming political in actions touching earthily down, dwelling nationally, and reaching virtually around a perilously entangled planet. Not to a once and for all success but to a better failure, all in all. No almighty fix, but some lure, glimmering darkly, even in lament.

TWISTED HOPE

Early in the poetic masterpiece composed in his name, we find Job in the full throes of his lamentation. "For the arrows of the

Almighty are in me; my spirit drinks their poison" (Job 7:11). He will not be repressed: "I will complain in the bitterness of my soul."(Job 6:4). Job only wants now for the God who has killed his children and ruined his body to finish the job: "That it would please God to crush me" (Job 6:9). His complaint is anguishingly personal. But its political implications rumble through. Already the deity of sovereign omnipotence was being exposed in His injustice. "Do not human beings have a hard service on earth, and are not their days like days of a laborer? Or like a slave who longs for the shadow?" (Job 7:1–2a).

Only after an epic drama of complaint, the proof text of every theodicy, every challenge to divine righteousness, does God appear in the climactic theopoetics of the whirlwind. And this whole epiphany unfolds in the form of questions back at Job, often misread as a bully's braggadocio: "Where were you when I laid the foundations of the earth?" (Job 38:4). The questioning continues in this cosmic vein for over a hundred verses. God answers, in other words, not with The Answer but with more questions. What manifests in the wild wind is a deity utterly entranced by every detail of the nonhuman universe. Even the glaciers: "From whose womb did the ice come forth, and who has given birth to the frost of heaven? The waters become hard like stone, and the face of the deep is frozen" (Job 38:29–30). And on the poem blows, detailing lightning, floods, lions, goats giving birth, calving deer; and through the comedy of the ostrich, with wings flapping wildly, forgetting its vulnerability to trampling, so that "when it spreads its plumes aloft, it laughs at the horse and its riders" (Job 39:18). God is laughing with, not at, the creaturely *chutzpah*.

The whirlwind voice is provoking Job to an altogether different theology, carrying a radical challenge to an already normalized anthropocentrism.[26] It is moved by the great Spirit of the universe, energetically involved in all that has been born, every creature—no exceptions. This ecogod offers no sentimental comfort, as it blows away Job's anthropocentric delusions. An ancient depiction of a cosmos brimming with creativity, unfurling at the edge of chaos, revealed in the reveling whirlwind, seems to whisk away the theopolitics of omnipotent and exceptionalist intervention. The opening and closing legend of God's bet with Satan and Job's reward appears then as an ironic framing device for a barely conceivable plenitude of life—of life amidst precarity—in which human life must find itself. Far from any punishment for this challenge to His sovereign decision, Job's furious lament is answered with the longest divine soliloquy in the Bible.[27] But what calls Job back, here, at *this* book's end, precedes the revelation: it is the precise poetics of his trauma. Early in his complaint, having described nights of sleepless torment, Job puts it like this:

> My days are swifter than a weaver's shuttle,
> And come to their end without hope [*tiqvah*].

In this speeding time in which one cannot breathe, rest, or live—no feeling of the now-moment—the poet evokes the sense of time running toward a meaningless end after unbearable loss. Time is running out. Job here becomes a parable of current ecopolitical catastrophe. But look at this language: the "weaver's shuttle," we recall from the Greek etymology of *kairos*, was an ancient image of technological hyperspeed. Most of us never see such skilled handweaving in action, the shuttle of the loom shooting like an arrow through the warp and woof of the textile. To be even swifter than the shuttle is to be skidding dazed and traumatized toward tragic shutdown.

To come to our end without hope—is this not the future we fear? Is it not already the future-present of whole populations of tormented humans, yoked in multiple earth temporalities to the nonhuman extinctions? Does this hopelessness not lead many politically responsible and ecologically sensitive thinkers to abandon hope as nothing but cruel optimism? For the sake of honesty then, I "speak in the anguish of my spirit" (Job 7:11).

"Without hope." The Hebrew word the poet has chosen for hope is *tiqvah*.[28] It means literally "a collection of fibers that are twisted together to make a strong and firm cord." It comes from the verb root meaning "to collect." So we may bind it with the sunestalemnos, collected or contracted, that provides the Pauline fiber of this political theology. This fibrous hope has nothing to do with a future abstracted from the present, with pictures of a heavenly happy ending or just a marketably better tomorrow. First Testament scholar Kenneth Ngwa calls Job's tiqvah a "trauma-hope."[29] Tiqvah names not a projection of future but an interweave of now; it signifies a cord "collected," made strong, by interweaving multiple fibers of time and narrative. And so the poem's dark brilliance juxtaposes a dissociated speed to the temporal rhythm of the shuttling interweave of life. It is a twisted hope that binds us to the loom of life now, in the now-moment that knots or nets the past and its traumatism to the possible. To the future not of an illusion but of the present. If we cannot hold it, bind it, in some fleshly mattering sense, it is not hope.

PRACTICE IN PROCESS

Spoken and unspoken, a political theology of the earth interweaves its tenses: an ancient history contracts in the dark hope of fresh assemblage. In this way, as is evident in the manifestly theological language of much of the present book, theology does not surrender once for all to secularization. But it may give itself—for all—to new secularreligious realizations. In the apophasis of improbable actualizations, the kairos turns up, once again, as now. Theology knows itself, in its apophatic action, to be theology. And so it casts its bread on the waters, chaotic though they be. What is it that "returns manifold"?

As exemplification of that apophatic wave-action between speech and action, I offer a case study close to the heart of the theology of the last chapter. If Whitehead provided the first explicit repudiation of divine exceptionalism, it is from his clues—divine lure, "poet of the universe"—that process theology constructed its attractive alternative. We considered how its God who offers possibility and internalizes consequences took the place of the all-controlling sovereign. The potential *novum* of each moment—usually trivial but sometimes revolutionary—marks the timely grace of the inception. Yet process theology developed not so much a negative as a constructive theology, an affirmation of alternative "metaphors mutely appealing." If some of us can contract its kataphasis to our apophasis, it is because, in the imaginative honesty of its unorthodox, open-ended world schema, it resists any fundamental certitude, including its own. This practiced humility permits the unsaying—precisely as political strategy—of its own theology. So it will be helpful to notice certain surprising self-secularizations of process theology.

Consider then the work of the leading process theologian, John B. Cobb Jr. For over half a century, he has written books in richly theological, avowedly Christian, language (dozens, at levels both lay and academic). But several of his texts are written in a rigorously secular vocabulary: from his early warning about climate change, *Is It Too Late?*, accompanied by his organizing of the first ecology and religion conference (1971); to *The Liberation of Life*, written with a biologist; to the magnum opus of ecological economics, *For the Common Good*, coauthored with an economist. All along, he has been at transdisciplinary pains to reach—as a theologian—beyond Christian or any religious discourse and at the same time beyond any enclosed discipline.[30] He speaks of "Ways," or "Wisdoms," rather than "world religions." And he reads the founders of the Ways (Moses, Laotzu, Confucius, Buddha, Jesus, Mohammed, etc.) not to be "religious" leaders but *secularizers*, insistently worldly, critical of cultic self-enclosure. Fast forward to Cobb's gathering the impressive public of an enormous conference, "Seizing an Alternative" (2015). Its multidisciplinary, multilateral effects continue to vibrate in the form of activist self-organization, as in the LA area where the Pando Populus network—taking its image from the single arboreal organism consisting of over one hundred acres of quaking aspen tree/s connected by one rhizome—works with a great variety of civic organizations and social movements, few of which are religion identified.[31] The publications multiply rhizomatically as well; indeed, during that same summer appeared a volume of sixty essays, including such leading secular activists as Bill McKibben and Vandana Shiva, in immediate response to Pope Francis' *Laudato Si'*, seeking to amplify its ecosocially theological impact.[32]

During this process epoch, in conjunction with the decades-long "China Project" of the Center for Process Studies, over two dozen centers for process thought in mainland China have sprung into existence. None of them employ any process-*theological* language, except occasional engagement of scholarship on Confucianism or Taoism. They are oriented to education for "ecological civilization." Cobb has returned several times to China, advocating a shift of agricultural policy away from the U.S. petrochemically pumped agribusiness model toward conservation and intensification of sustainable rural practices. His language is then strictly secular—and so exemplifies the apophatic action of an earthen political theology. And its effect? This is impossible to calculate, but he has been reportedy quoted approvingly by the premier of China and given speaking opportunities at high levels of government. His close colleague, the theologian Philip Clayton, considers his own viewpoint "Hindu-Christian." Clayton published *Organic Marxism*—free of any religious language—with its Chinese translation in mind.[33] Within a socialist lexicon resonant with the history of the People's Republic, it presses for commitment to a sustainable ecology. It became a fast best seller assigned in hundreds of the required Marxism courses of the land. Dozens of dissertations and over seventy articles have been written on it so far. That a coordinated attack against the process impact of Cobb and Clayton (accusing it of pursuing a secret Christian mission) has not so far managed to stifle their influence is evidence that it is "failing better" than one would have any reason to predict. And yet Premier Xi recently absolutized his sovereignty, making his exception to the Chinese constitution its revised rule. This move, in line with the wave of global strongmen, cancels nothing of the struggle but adds to its agonism.

Conspiring with Cobb, Clayton has organized the network EcoCiv. It has launched projects in Africa, Asia, Europe as well as the U.S. "EcoCiv propels scholars, activists, and policy experts toward realizing an ecological civilization—a fully sustainable human society in harmony with surrounding ecosystems and communities of life. We provide momentum for integrating work across sectors that leads to concrete action, including grassroots innovations, policy reform, and laying the foundations for a sustainable future." Here, its mission statement exemplifies the theological apophasis of action, which helps it to gather a radically diverse public: "Together, we seek to identify how social, political, and economic life needs to be organized if humanity is to achieve a sustainable, ecological society over the long-term."[34]

Such practices refuse to let the capital ugliness of their native lands unhinge local planetary struggles. EcoCiv contracts in its struggle a lively potentiality. How probable? Spoken or unspoken depending on the context of collaboration, the theology here in play activates improbable possibilities. It has a deep hunch. It does not calculate the odds of success, let alone grant some providential guarantee of a good outcome. *On the contrary*, It was after all Cobb who asked vis-à-vis early evidence of systemic collapse of the planet's ecosystems, "Is it too late?" His response to that question, now or then, is a repudiation of false determinisms and a call to collective responsibility.

Such counterexceptionalist theology does not work in the panic of emergency but the process of emergence. Having deconstructed the dictatorship of outcomes, it lives gracefully with its unknowing. It teaches a rich enjoyment of the process itself, of the animated embodiment of all "actual occasions of experience." As traced earlier, Muñoz' "sensuous intersectionality" resonates with the physical interrelating of the becoming occasions. Similarly process theology fosters a political pluralism of interconnected differences. Thus, along a different but intersecting axis, the praxis of womanist process theologian Monica Coleman explores a nexus of race, sex, and trauma; her extraordinary transcription of bipolar disorder undoes at once the normativity and the exceptionalism of any sovereign subject.[35]

Attending to the lure of the moment, human creatures occasionally do actually rise to the occasion. In the very midst of ungodly obstructions and unbearable damages, personal, political, planetary, any moment may shuttle kairotic potentiality. No almighty fix, but some lure, glimmering darkly, even in lament. In radical fidelity to that lure, Job's drama had performed an arc from a chronos hopelessly running out to a kairos of the creatures of the mattering multitemporal universe. Its tiqvah puts me in mind of the last line of the EcoCiv mission statement: "Where hopelessness arises, we call others to join us in walking toward ecological civilization, one step at a time." Each act of solidarity may spring open some subtle possibility.

The possible future looms—it does not conclude, stay put, or guarantee. It remains a modality of the quivering present, capable of collecting us in some fresh interweave of becomings. Even amidst catastrophe. It may take a whirlwind as catalyst. In the loom of theopoetics, without a God to blame or to beg, the politics of theology and the theology of politics twist into a dense fabric of public response-ability.

It is the chance of ecosocial inception, the emergence of a new public and its new earth, that a political theology of the earth nurtures. As theology, it cannot look forward without collecting the past of a messianic potentiality. Indeed, we can only conclude now by recollecting a particular theological secularization of the messianic now-time. In the context of earlier U.S. emergency, it yielded the best rendition of the Pauline kairos ever: "the fierce urgency of now." Thus King in his Riverside sermon also updated ancient prophetic warning: "There is such a thing as being too late."[36] Emergency is dis/closed as the urgent chance of an improbable emergence. This was no more a declaration of The End than was Cobb's urgently inquisitive "Is it too late?" To decide that it *is* too late—that would now be as irresponsible as living like there is no such thing. Opposite modes, but either way, we provide ourselves an alibi, an excuse, an exit from the troubles. To stay with the struggle means to enter not the continuum of dread ("I do not want you to be anxious") but the wake of mourning, the energy of indeterminacy, and the awakening potential of this now. What matters unconditionally may materialize under the most urgent conditions.

The dark space of possibility opens along the edge of uncertainty, flickering between the apocalyptic and the apophatic. Sometimes the flicker is a flutter, a flock, a flight: "A Murmuration of Starlings! Gorgeous, mysterious, no one in the lead, everyone in the lead . . ." The message comes to me this morning, from a longtime collaborator in process-theological practices, sharing out of the blue a burst of new zoology. It attends to the high intelligence of birds. I cannot repress the synchronicity delivering this parable of now:

> Each one is managing "uncertainty in sensing" by attending to seven other birds, thus optimizing "the balance between group

cohesiveness and individual effort." There is probably more to understand.... Which seven do they pick and why? Proximity, pheromones, sparkle in a feather? Among one bird's seven, how many of those are attending to at least some not in the same seven as the first bird? What is happening when the seven becomes part of 700? What tips them to and fro as they go hither and yon?[37]

Do those questions faintly echo the cosmic animacy of the voice from the whirlwind? Might we in animal apophasis mimic this collectivity, collect ourselves, collect each other, in a dance of no one and everyone in the lead? In this counterexceptionalist murmuration, pulsing with interwoven difference, do we get a hunch as to our own political possibilities? Black-draped as we come, we have it in us to get it together. In the fierce urgency of our all too human now, what local planetary solidarity might emerge? Never quite all in all, and yet nonetheless—all in. Why not become the new earth, the new public, we imagine?

NOTES

BEGINNING

1. Giorgio Agamben, *The Time That Remains: A Commentary on the Letter to the Romans*, trans. by Patricia Dailey (Stanford, CA: Stanford University Press, 2005).
2. Kristina Zolatova, "Eschatological Developments Within the Pauline Corpus," in *Per Caritatem*, 4 February 2010, http://percaritatem.com/2010/02/04/eschatological-developments-within-the-pauline-corpus/.
3. Paul Tillich, *Systematic Theology*, vol. 3: *Life and the Spirit/History and the Kingdom of God* (Chicago: University of Chicago Press, 1976 [1963]), 369:

    > This term has been frequently used since we introduced it into theological and philosophical discussion in connection with the religious socialist movement in Germany after the First World War. It was chosen to remind Christian theology of the fact that the biblical writers, not only of the Old but also of the New Testament, were aware of the self-transcending dynamics of history. And it was chosen to remind philosophy of the necessity of dealing with history, not in terms of its logical and categorical structure only, but also in terms of its dynamics.... Its original meaning—the right time, the time in which something can be done—must be contrasted with *chronos*, measured time or clock time. The former is qualitative, the latter quantitative.

4. Tillich, *Systematic Theology*, 3:369: "We spoke of the moment at which history, in terms of a concrete situation, had matured to the point of being able to receive the breakthrough of the central manifestation of the Kingdom of God. The New Testament has called this moment the 'fulfilment of time,' in Greek, *kairos*." My thanks to John Thatamanil for reconnecting me to Tillich.
5. Eric Charles White, *Kaironomia: On the Will-to-Invent* (Ithaca, NY: Cornell University Press, 1987), 13.
6. Kairos Theologians, *The Kairos Document: Challenge to the Church, Theological Comment on the Political Crisis in South Africa (25 September 1985)* (Grand Rapids, MI: Eerdmans, 1986).
7. Agamben, *The Time That Remains*, 19 (my emphasis).
8. Walter Benjamin, "Theses on the Philosophy of History," in *Illuminations: Essays and Reflections*, ed. Hannah Arendt, trans. Harry Zohn (New York: Schocken, 2007 [1968]), 253–264. See also Daniel Weidner, "Thinking beyond Secularization: Walter Benjamin, The 'Religious Turn,' and the Poetics of Theory," *New German Critique* 37, no. 3 (2010): http://www.zfl-berlin.org/tl_files/zfl/downloads/personen/weidner/Benjamin_secularization.pdf.
9. L. L. Welborn, *Paul's Summons to Messianic Life: Political Theology and the Coming Awakening* (New York: Columbia University Press, 2015), 20.
10. Welborn, *Paul's Summons to Messianic Life*, 16.
11. Agamben, *The Time That Remains*, 64.
12. On the resemblance of the Pauline kairos to the gospel teaching of the kingdom, Welborn writes: "The similarity of the Jesus-tradition preserved in Mark 1:15 and Luke 10:9–10 with Romans 13:11–12 suggests that Paul's concept of the nun kairos presupposes Jesus's proclamation of the kingdom of God. Thus, for Paul, as for Jesus, the kairos is not an interval before the end of time, but time filled with the presence of the 'now.'" Welborn, *Paul's Summons to Messianic Life*, 20.
13. I hear here the hymn of James K. Manley, "Spirit, Spirit of Gentleness" (1978): "You call from tomorrow / you break ancient schemes, / from the bondage of sorrow the captives dream dreams, / our women see visions, our men clear their eyes, / with bold new decisions your people arise."

14. See, for example, Peter Scott, *A Political Theology of Nature* (New York: Cambridge University Press, 2003); and Michael S. Northcott, *A Political Theology of Climate Change* (Grand Rapids, MI: Eerdmans, 2013).
15. Nicholas of Cusa, "On Learned Ignorance," in *Selected Spiritual Writings*, trans. H. Lawrence Bond (Mahwah, NJ: Paulist Press, 1997), 140 [II.5.118]; See also my work on Cusa's mystical cosmology in Catherine Keller, *Cloud of the Impossible: Negative Theology and Planetary Entanglement* (New York: Columbia University Press, 2014), esp. chap. 3, "Enfolding and Unfolding God: Cusanic *Complicatio*."
16. We will consider further in chapters 2 and 3 this notion of the actual occasion, Whitehead's term for the elemental units of the world: "The final facts are, all alike, actual entities, and these actual entities are drops of experience, complex and interdependent." Alfred North Whitehead, *Process and Reality: An Essay in Cosmology*, corrected ed. (New York: Free Press, 1973 [1929]), 18.
17. See Namsoon Kang, *Cosmopolitan Theology: Reconstituting Planetary Hospitality, Neighbor-Love, and Solidarity in an Uneven World* (St. Louis, MO: Chalice, 2013); and Daniele Archibugi, ed., *Debating Cosmopolitics* (New York: Verso, 2003); and, for her superb rendition of "the kingdom of God," see Dhawn Martin, "The Cosmopolis of God: A Political Theology of the Kingdom" (Ph.D. diss., Drew University).
18. Mike Davis, *Planet of Slums* (New York: Verso, 2006).
19. Carl Schmitt, *Political Theology: Four Chapters on the Concept of Sovereignty*, trans. by George Schwab (Chicago: University of Chicago Press, 2005), 36.
20. Schmitt, *Political Theology*, 36.
21. Mikhail Bakunin, "The Political Theology of Mazzini and the International," trans. Sarah E. Holmes, available via Libertarian-Labyrinth Wiki, http://wiki.libertarian-labyrinth.org/index.php?title=The_Political_Theology_of_Mazzini_and_the_International.
22. Schmitt, *Political Theology*, 66.
23. For a helpful history of the Black Social Gospel movement in particular, see Gary Dorrien, *The New Abolition: W. E. B. Du Bois and the Black Social Gospel* (New Haven: Yale University Press, 2015).

24. See John Thatamanil, *Theology Without Borders: Religious Diversity and Theological Method* (New York: Fordham University Press, forthcoming).
25. For a survey of Metz's, Moltmann's, and Sölle's respective engagements of political theology in the mid-twentieth century, see John B. Cobb Jr.'s *Process Theology as Political Theology* (Philadelphia: Westminster, 1982), esp. chap. 1, "The Challenge of Political Theology."
26. Jürgen Moltmann, *The Living God and the Fullness of Life*, trans. Margaret Kohl (Louisville, KY: Westminster John Knox, 2015).
27. See Clayton Crockett and Jeffrey W. Robbins, in particular their edited volume with Ward Blanton and Noëlle Vahanian, *An Insurrectionist Manifesto: Four New Gospels for a Radical Politics* (New York: Columbia University Press, 2016). See also Clayton Crockett, *Radical Political Theology: Religion and Politics After Liberalism* (New York: Columbia University Press, 2013); and Jeffrey W. Robbins, *Radical Democracy and Political Theology* (New York: Columbia University Press, 2011).
28. Robbins, *Radical Democracy and Political Theology*, 191.
29. See especially Northcott's reflections on Schmitt's *The Nomos of the Earth, The Nomos of the Earth in the International Law of Jus Publicum Europaeum*, trans. G. L. Ulmen (Candor, NY: Telos, 2006 [1950]), in his *A Political Theology of Climate Change*, chap. 3: "The *Nomos* of the Earth and Governing the Anthropocene." For the first major text of this genre of ecologically driven political theology, see the impressive trinitarian doctrine of creation of Peter Scott, *A Political Theology of Nature* (Cambridge: Cambridge University Press), 2003. See also texts rooting this genre in the U.S. situation: Michael S. Hogue, *American Immanence: Democracy for an Uncertain World* (New York: Columbia University Press, 2018); and S. Yael Dennis, *Edible Entanglements: On a Political Theology of Food* (Eugene, OR: Cascade, 2018).
30. Augustine, *The City of God*, trans. Macus Dods, D.D., intro. Thomas Merton (New York: Modern Library, 1950); for a powerful analysis of the political theology of Augustine's two cities as well as Calvin's version, see Martin, "The Cosmopolis of God."
31. The classic of ecofeminist theology is Rosemary Ruether's *Gaia and God: An Ecofeminist Theology of Earth Healing* (New York: HarperCollins, 1992). Among key twenty-first-century texts, see Cynthia D.

Moe-Lobeda, *Resisting Structural Evil: Love as Ecological-Economic Vocation* (Minneapolis: Fortress, 2013); Melanie L. Harris, *Ecowomanism: African American Women and Earth-Honoring Faiths* (Maryknoll, NY: Orbis, 2017); and Grace Ji-Sun Kim and Hilda P. Koster, eds., *Planetary Solidarity: Global Women's Voices on Christian Doctrine and Climate Justice* (Minneapolis: Fortress, 2017).

32. See, for example, John B. Cobb Jr., *Is It Too Late? A Theology of Ecology*, rev. ed. (Denton, TX: Environmental Ethics, 1995 [1971]); and also his work with the economist Herman E. Daly, *For the Common Good: Redirecting the Economy Toward Community, the Environment, and a Sustainable Future*, 2d ed. (Boston: Beacon, 1994 [1989]).

33. See John B. Cobb Jr., *Spiritual Bankruptcy: A Prophetic Call to Action* (Nashville: Abingdon, 2010).

34. William E. Connolly, *Aspirational Fascism: The Struggle for Multifaceted Democracy Under Trumpism* (Minneapolis: University of Minnesota Press, 2017).

35. See the activist and writer Rebecca Solnit's *Hope in the Dark: Untold Histories, Wild Possibilities*, 3d ed. (Chicago: Haymarket, 2016 [2004]).

36. For an account of the geopolitical crises that attend climate destabilization, see Christian Parenti, *Tropic of Chaos: Climate Change and the New Geography of Violence* (New York: Nation, 2011).

37. Judith Butler, "Walter Benjamin and the Critique of Violence," in *Parting Ways: Jewishness and the Critique of Zionism* (New York: Columbia University Press, 2012), 91.

38. Stefano Harney and Fred Moten, *The Undercommons: Fugitive Planning and Black Study* (New York: Minor Compositions, 2013).

I. POLITICAL

1. Antonio Spadaro and Marcelo Figueroa, "Evangelical Fundamentalism and Catholic Integralism: A Surprising Ecumenism," *La Civiltà Cattolica*, 13 July 2017, http://www.laciviltacattolica.it/articolo/evangelical-fundamentalism-and-catholic-inegralism-in-the-usa-a-surprising-ecumenism/.

2. See the papal encyclical from 2015, "'*Laudato Si'*: On Care for Our Common Home," originally delivered in Rome, 24 May 2015, http://

w2.vatican.va/content/francesco/en/encyclicals/documents/papa-francesco_20150524_enciclica-laudato-si.html.
3. For two persuasively refreshed concepts of "commonwealth" in the U.S. context, each theologically grounded, see David Ray Griffin, John B. Cobb Jr., Richard A. Falk, and Catherine Keller, *The American Empire and the Commonwealth of God: A Political, Economic, Religious Statement* (Louisville, KY: Westminster John Knox, 2006); and, with a savvy and engaging theological spin, tuned to the U.S. present, see Marcia Pally, *Commonwealth and Covenant: Economics, Politics, and Theologies of Relationality* (Grand Rapids, MI: Eerdmans, 2016).
4. For an abbreviated analysis of that very contraction in Paul's epistles, as it is elaborated by Giorgio Agamben, see discussion in the introduction, this volume.
5. Carl Schmitt, *The Concept of the Political*, expanded ed., trans. George Schwab (Chicago: University of Chicago Press, 2007 [1996]), 26.
6. Chantal Mouffe, *The Democratic Parodox* (New York: Verso, 2005 [2000]), 54.
7. Schmitt, *The Concept of the Political*, 29.
8. Schmitt, *The Concept of the Political*, 29.
9. I think not only of dynamics in my own turbulent family of origin but of the way a basic "we" such as whiteness is originated in the family we versus the racialized Other. See Thandeka, *Learning to Be White: Money, Race, and God in America* (New York: Bloomsbury, 2013 [1999]).
10. Carl Schmitt, "Theory of the Partisan: Intermediate Commentary on the Concept of the Political (1963)," *Telos* 127 (2004): 85.
11. For a look at vibrant activism toward the "birthing of a future where love is a public ethic" through interfaith engagement, see the website of the Revolutionary Love Project, an initiative out of the University of Southern California: http://www.revolutionarylove.net/.
12. Schmitt, *The Concept of the Political*, 29.
13. Chantal Mouffe, *The Return of the Political* (New York: Verso, 2005), 6.
14. As to the relationship of Mouffe's and Connolly's quite congruent concepts of "agonism," they seem to have developed along divergently parallel pathways. I thank Wren Hillis for the following account: "Connolly and Mouffe both rely on the work of Adorno and Foucault, and they also have some crossover sources they like to interrogate/

critique (i.e., Schmitt, Habermas, Rawls). However, behind these commonalities, Connolly is driven much more forcefully by a Nietzschean paradigm, whereas Mouffe is dealing with a more militantly post-Marxist radicalization using sources like Gramsci. Mouffe also critiques, without mentioning Connolly by name, those who operate out of an "ethical-particularist approach" whose vocabularies derive from "Levinas, Arendt, Heidegger or even Nietzche" (*The Democratic Paradox*, 129). One of the subtle differences this evokes is Connolly's rich attention to existential suffering and resentment/ressentiment, which appears to be less present in Mouffe's work. Like others, Mouffe's agonism recognizes that consensus is not the ideal or norm of society; struggle is inherent to the democratic process." Personal email, 19 February 2018.

15. William E. Connolly, *The Fragility of Things: Self-Organizing Processes, Neoliberal Fantasies, and Democratic Activism* (Durham, NC: Duke University Press, 2013), 133.
16. See William E. Connolly, *Capitalism and Christianity, American Style* (Durham, NC: Duke University Press, 2008), x.
17. Connolly, *Capitalism and Christianity*, 49.
18. "I do not suggest that faith in a transcendent Being automatically expresses an ethos of ressentiment and revenge, nor that this ethos is intrinsic to Christianity per se, nor that it is intrinsic to every version of evangelism." Connolly, *Capitalism and Christianity*, 52.
19. See my essay, "Foxangelicals, Political Theology, and Friends," in *Doing Theology in the Age of Trump: A Critical Report on the Threat of Christian Nationalism*, ed. Jeffrey W. Robbins and Clayton Crockett (Eugene, OR: Wipf and Stock, 2018).
20. Connolly, *The Fragility of Things*, 171.
21. Connolly, *Capitalism and Christianity*, 25.
22. For a deeply political theology of the activist hope that lament makes possible, see Emmanuel Katongole, *Born from Lament: The Theology and Politics of Hope in Africa* (Grand Rapids, MI: Eerdmans, 2017). See also his *The Sacrifice of Africa: A Political Theology of Africa* (Grand Rapids, MI: Eerdmanns, 2013).
23. Katongole, *Born from Lament*, 60.
24. Katongole, *Born from Lament*,

25. See my response, among many, to the encyclical *Laudato Si'*, "Encycling: One Feminist Theological Response," in *For Our Common Home: Process-Relational Reponses to Laudato Si*, ed. John B. Cobb Jr. and Ignacio Castuera (Claremont, CA: Process Century, 2015).
26. Stefano Harney and Fred Moten, *The Undercommons: Fugitive Planning and Black Study* (New York: Minor Compositions, 2013), 20.
27. Harney and Moten, *The Undercommons*, 20.
28. Harney and Moten, *The Undercommons*, 104.
29. Harney and Moten, *The Undercommons*, 20.
30. Emilie M. Townes, "If You Quare It You Can Change It: Changing the Boxes That Bind Us," in *Unsettling Science and Religion: Contributions and Questions from Queer Studies*, ed. Lisa Stenmark and Whitney Bauman (Lanham, MD: Lexington, 2018), 64.
31. Harney and Moten, *The Undercommons*, 20.
32. Harney and Moten, *The Undercommons*, 151.
33. Joseph R. Winters, *Hope Draped in Black: Race, Melancholy, and the Agony of Progress* (Durham, NC: Duke University Press, 2016). Thanks to my research assistant Winfield Goodwin for this find.
34. Philip Clayton integrates the political hope and the ecological potential of the socialist ideal in *Organic Marxism: An Alternative to Capitalism and Ecological Catastrophe*, with Justin Heinzekehr (Claremont, CA: Process Century, 2014).
35. See my *Face of the Deep: A Theology of Becoming* (New York: Routledge: 2003), esp. chap. 12, "*Docta ignorantia:* darkness on the face," which links the dynamics of racialization to the hermeneutics of creation.
36. Ilya Prigogine, cited in William E. Connolly, *A World of Becoming* (Durham, NC: Duke University Press, 2011), 96.
37. Connolly, *The Fragility of Things*, 6.
38. Connolly, *The Fragility of Things*, 190.
39. Karl Polanyi, *The Great Transformation: The Political and Economic Origins of Our Time* (Boston: Beacon, 2001 [1944]), 245.
40. See William E. Connolly, "Donald Trump and the New Fascism," The Contemporary Condition, August 2016, http://contemporarycondition.blogspot.com/2016/08/donald-trump-and-new-fascism.html.
41. See William E. Connolly, *Aspirational Fascism: The Struggle for Multifaceted Democracy Under Trumpism* (Minneapolis: University of Minnesota Press, 2017).

1. POLITICAL ൞ 189

42. Connolly, *The Fragility of Things*, 21.
43. Connolly, *The Fragility of Things*, 25.
44. See Connolly's reading of Nietzsche and Whitehead through one another in *The Fragility of Things*, chap. 4, "Process Philosophy and Planetary Politics."
45. Connolly, *The Fragility of Things*, 9.
46. Connolly, *Aspirational Fascism*, 92.
47. Timothy Morton, *Hyperobjects: Philosophy and Ecology After the End of the World* (Minneapolis: University of Minnesota Press, 2013).
48. *Everywhere and Forever: Mahler's Song of the Earth*, a film by Jason Starr (Pleasantville, NY: Video Artists International, 2014), DVD4585.
49. Connolly, *The Fragility of Things*, 19.
50. Connolly, *The Fragility of Things*, 193–94.
51. See Moe-Lobeda's "Climate Debt, White Privilege, and Christian Ethics as Political Theology," in *Common Goods: Economy, Ecology, Political Theology*, ed. Catherine Keller, Melanie Johnson-DeBaufre, and Elias Ortega-Aponte (New York: Fordham University Press, 2015).
52. Carl Schmitt, *Political Theology: Four Chapters on the Concept of Sovereignty*, trans. George Schwab (Chicago: University of Chicago Press, 2005 [1985]), 5.
53. Schmitt, *Political Theology*, 36.
54. Schmitt, *Political Theology*, 36.
55. Schmitt, *Political Theology*, 15.
56. Søren Kierkegaard, "Concluding Letter, Repetition," in *Fear and Trembling / Repetition*, ed. and trans. Howard V. Hong and Edna H. Hong (Princeton: Princeton University Press, 1983), 227.
57. Kierkegaard, "Concluding Letter, Repetition," 227.
58. Kathryn Tanner, *Economy of Grace* (Minneapolis: Fortress, 2005), 142.
59. Donald J. Trump, Twitter post, 13 October 2015, 2:43 am, https://twitter.com/realdonaldtrump/status/653868764094722048?lang=en.
60. Schmitt, *Political Theology*, 36.
61. Schmitt, *Political Theology*, 64.
62. Schmitt, *Political Theology*, 66.
63. See Keller, *Face of the Deep*, 74–75.
64. "Yet there was not absolutely nothing. There was a certain formlessness devoid of any specific character" (*Confessions* XII.3). *The Confessions of Saint Augustine*, trans. John K. Ryan (Garden City, NY:

Image, 1960), 306. When Augustine fixed his thoughts "on the bodies themselves, and poured more deeply into their mutability," did he discern a "nothingsomething" (*nihil aliquid*) (XII.6). Cited in Keller, *Face of the Deep*, 74–75. See also my article with Virginia Burrus, "Confessing Monica: Reading Augustine Reading His Mother," reprinted in Catherine Keller, *Intercarnations: Exercises in Theological Possibility* (New York: Fordham University Press, 2017).

65. See Keller, *Face of the Deep*, chap. 6: "Sea of Heteroglossia: return of the biblical chaos."

66. *Elohim*, a word for "God" in Hebrew, is a singular noun with a plural ending.

67. The postsecular is especially associated with Charles Taylor, *A Secular Age* (Cambridge, MA: Belknap Press of Harvard University Press, 2007); Jürgen Habermas, "Secularism's Crisis of Faith: Notes on Post-Secular Society," *New Perspectives Quarterly* 25 (2008): 17–29; and Rosi Braidotti, "In Spite of the Times: The Postsecular Turn in Feminism," *Theory, Culture, and Society* 25 (2008): 1–24.

68. See José Casanova, "Rethinking Secularization: A Global Comparative Perspective," in *Religion, Globalization, and Culture,* ed. Peter Beyer and Lori Beaman (Leiden: Koninklijke Brill NV, 2007), originally published in the *Hedgehog Review*'s double issue, "After Secularization," vol. 8, nos. 1–2 (2006): 7–22.

69. Eusebius of Caesaria, "From a Speech for the Thirtieth Anniversary of Constantine's Accession," in *From Irenaeus to Grotius: A Sourcebook in Christian Political Thought*, ed. Oliver O'Donovan and Joan Lockwood O'Donovan (Grand Rapids, MI: Eerdmans, 1999), 60. See also Eusebius, "From a Speech on the Dedication of the Holy Sepulchre Church," in *From Irenaeus to Grotius,* 60:

> For before this time the various countries of the world, as Syria, Asia, Macedonia, Egypt and Arabia, had been severally subject to different rules. The Jewish people, again, had established their dominion in the land of Palestine. And these nations, in every village, city and district, actuated by the same insane spirit, were emerged in incessant and murderous war and conflict. But two mighty powers starting from the same point, the Roman empire

which henceforth was swayed by a single sovereign, and the Christian religion, subdued and reconciled these contending elements. Our Savior's mighty power destroyed at once the many governments and the many gods of the powers of darkness, and proclaimed to all men, both rude and civilized, to the extremities of the earth, the sole sovereignty of God himself.

70. Kelly Brown Douglas, *Stand Your Ground: Black Bodies and the Justice of God* (Maryknoll, NY: Orbis, 2015), 7.
71. Brown Douglas, *Stand Your Ground*, 5.
72. Brown Douglas, *Stand Your Ground*, 8.
73. Catherine Keller, *Apocalypse Now and Then: A Feminist Guide to the End of the World* (Boston: Beacon, 1996), 38f, 116, 136.
74. Benjamin Franklin, "Observations Concerning the Increase of Mankind, Peopling of Countries, etc.," cited in Brown Douglas, *Stand Your Ground*, 17.
75. Brown Douglas, *Stand Your Ground*, 15.
76. Brown Douglas, *Stand Your Ground*, 28, 39.
77. Carol Wayne White, *Black Lives and Sacred Humanity: Toward an African American Religious Naturalism* (New York: Fordham University Press, 2016), 33.
78. J. Kameron Carter, "Between W. E. B. Du Bois and Karl Barth: The Problem of Modern Political Theology," in *Race and Political Theology*, ed. Vincent W. Lloyd (Stanford, CA: Stanford University Press, 2012), 89.
79. Giorgio Agamben, *The Time That Remains: A Commentary on the Letter to the Romans*, trans. by Patricia Dailey (Stanford, CA: Stanford University Press, 2005), 108.
80. Giorgio Agamben, *State of Exception*, trans. Kevin Attell (Chicago: University of Chicago Press, 2005), 6–7.
81. Agamben, *State of Exception*, 87.
82. Agamben, *State of Exception*, 87.
83. Walter Benjamin, cited in Agamben, *State of Exception*, 6.
84. Agamben, *State of Exception*, 87.
85. Quinta Jurecic, "Donald Trump's State of Exception," *Lawfare*, 14 December 2016, https://lawfareblog.com/donald-trumps-state-exception.

86. Max Weber, "Charismatic Authority," in *The Theory of Social and Economic Organization*, trans. A. R. Anderson and Talcott Parsons (New York: Free Press, 1964), 358–63 (my emphasis).
87. Walter Benjamin, "Theses on the Philosophy of History," in *Illuminations: Essays and Reflections*, ed. by Hannah Arendt, trans. Harry Zohn (New York: Schocken, 2007 [1969]), 254.
88. Benjamin, "Theses on the Philosophy of History," 257.
89. For a rich analysis of the divine "passion" as an eros that in "groans and birthpangs" at once suffers and enjoys the cosmos, see the decolonial theology of Elaine Padilla, *Divine Enjoyment: A Theology of Passion and Exuberance* (New York: Fordham University Press, 2015).
90. Ernst Bloch, *Atheism in Christianity: The Religion of the Exodus and the Kingdom*, trans. J. T. Swann (New York: Verso, 2009 [1972]).
91. See Keller, *Apocalypse Now and Then*, 122.
92. José Esteban Muñoz, *Cruising Utopia: The Then and There of Queer Futurity* (New York: New York University Press, 2009), 92.
93. Muñoz, *Cruising Utopia*, 95.
94. Muñoz, *Cruising Utopia*, 96.
95. In *Terrorist Assemblages*, Jasbir Puar picks up here: "In the state of exception the exception insidiously becomes the rule. . . . Sexual exceptionalism also works by glossing over its policing of the boundaries of acceptable gender, racial and class formations." So the mobilizing of the exceptional status of the homosexual within the United States and some European nations generates what she calls "homonationalism": the exceptional, acceptable, properly Islamophobic—usually white and affluent—homosexual produced in opposition at once to "the colloquial deployment of Islamic sexual repression that plagues human rights, liberal, queer, and feminist discourses, and the Orientalist wet dreams of lascivious excesses of pedophilia, sodomy, and perverse sexuality." Puar, *Terrorist Assemblages: Homonationalism in Queer Times* (Durham, NC: Duke University Press, 2007), 14.
96. "For our knowledge and comprehension of reality, and our reflections on it, that means at least this: that in the medium of hope our theological concepts become not judgments which nail reality down to what it is, but anticipations which show reality its prospects and its future possibilities." Jürgen Moltmann, *Theology of Hope: On the Ground and*

I. POLITICAL ⌘ 193

Implications of a Christian Eschatology, trans. Margaret Kohl (Minneapolis: Fortress, 1993 [1967]), 35.
97. Jürgen Moltmann, *The Living God and the Fullness of Life*, trans. Margaret Kohl (Louisville, KY: Westminster John Knox Press, 2015), 180.
98. Data available in the last two decades from Eastern Europe has corrected the widespread assumption that Stalin was responsible for more noncombatant deaths than Hitler. See Timothy Snyder, "Hitler or Stalin: Who Killed More?," http://www.nybooks.com/articles/2011/03/10/hitler-vs-stalin-who-killed-more/.
99. Thanks to my research assistant Winfield Goodwin for bringing to my attention Whitehead's overt influence on Bloch.
100. Keller, *Apocalypse Now and Then*, 122.
101. Paulina Ochoa Espejo, *The Time of Popular Sovereignty: Process and the Democratic State* (University Park: Pennsylvania State University Press, 2011). See also S. Yael Dennis, *Edible Entanglements: On a Political Theology of Food* (Eugene, OR: Cascade, 2018).
102. Benjamin, "Theses on the Philosophy of History," 261.
103. Benjamin, "Theses on the Philosophy of History," 264; 263.
104. Carl Schmitt, cited in Agamben, *The Time That Remains*, 110.
105. Jacob Taubes, *The Political Theology of Paul*, trans. Dana Hollander (Stanford, CA: Stanford University Press, 2003), 69.
106. Especially my erstwhile colleague the mujerista theologian Ada Maria Isasi Diaz deployed "kindom of God" as the emancipatory translation of *basileia theou*.
107. Clayton Crockett, "Earth: What Can a Planet Do?" in *An Insurrectionist Manifesto: Four New Gospels for a Radical Politics*, ed. Ward Blanton, Clayton Crockett, Jeffrey Robbins, and Noëlle Vahanian (New York: Columbia University Press, 2016), 57.
108. Connolly, *Aspirational Fascism*, 93.
109. "The planet is in the species of alterity, belonging to another system, and yet we inhabit it on loan." Gayatri Spivak, *Death of a Discipline* (New York: Columbia University Press, 2003), 72. See also *Planetary Loves: Spivak, Postcoloniality, and Theology*, ed. Stephen D. Moore and Mayra Rivera (New York: Fordham University Press, 2011).
110. Michael S. Northcott, *Place, Ecology, and the Sacred: The Moral Geography of Sustainable Communities* (New York: Bloomsbury, 2015), 175.

111. Saba Mahmood, *Politics of Piety: The Islamic Revival and the Feminist Subject* (Princeton: Princeton University Press, 2005). See also Keller, "The Queer Multiplicity of Becoming," in *Intercarnations*.
112. "From the position of Christian theology, the secular is the place where the religious is still always present, but implicitly so.... Modest, implicit, and without boasting, this is still very much a spirituality recognizing the experience of God's rootedness in all life." Trygve Wyller, "The Discovery of the Secular-Religious Other," in *Reformation Theology for a Post-Secular Age: Logstrup, Prenter, Wingren, and the Future of Scandinavian Creation Theology*, ed. Neils Henrik Gregersen, Bengt Kristensson Uggla, and Trygve Wyller (Göttingen: Vandenhoeck & Ruprecht, 2017), 255.
113. Crockett, "Earth," 57.
114. The theologian then adds, "Kairos time is God's time." Brown Douglas, *Standing Your Ground*, 206. This author says "amen."

2. EARTH

1. Ed Roberson, "To See the Earth Before the End of the World," in *To See the Earth Before the End of the World* (Middletown, CT: Wesleyan University Press, 2011), 3.
2. Roberson, "To See the Earth Before the End of the World," 3.
3. Daniel Fagre quoted in Andrea Thompson, "Glacier National Park Is Losing Its Glaciers," Climate Central, 10 May 2017, http://www.climatecentral.org/news/glacier-national-park-losing-its-glaciers-21436.
4. Christopher White, *The Melting World: A Journey Across America's Vanishing Glaciers* (New York: St. Martin's, 2013), 228.
5. "That warming is happening faster in western Montana, where temperatures have increased at a rate almost double the global average," Fagre explains to Montana Public Radio: "We are kind of a bellwether, we're an early indicator of the kinds of changes that are going to occur elsewhere." Quoted in Merrit Kennedy, "Disappearing Montana Glaciers a 'Bellwether' Of Melting to Come?," *NPR*, 11 May 2017, http://www.npr.org/sections/thetwo-way/2017/05/11/527941678/disappearing-montana-glaciers-a-bellwether-of-melting-to-come.

Bellwethers and apocalypses: a recent study by the journal *Paleoworld* warns of apocalyptic consequences of glacial melting and methane release in the Arctic; see Darh Jamail, "Release of Arctic Methane 'May Be Apocalyptic,' Study Warns," *Truthout*, 23 March 2017, http://www.truth-out.org/news/item/39957-release-of-arctic-methane-may-be-apocalyptic-study-warns.

6. See the *New York Times* multimedia piece on this Alaskan melting in Henry Fountain, "Alaska's Permafrost Is Thawing," *New York Times*, 23 August 2017.
7. Roberson, "To See the Earth Before the End of the World," 3.
8. Cf. the early section "Kairos and Contraction" of the introduction, "A Beginning," in this volume.
9. Coral Davenport, "With Trump in Charge, Climate Change References Purged From Website," *New York Times*, 20 January 2017, https://www.nytimes.com/2017/01/20/us/politics/trump-white-house-website.html. Commitments to multiple other vulnerable groups, notably LGBTQ rights, also quickly disappeared.
10. See Christian Parenti, *Tropic of Chaos: Climate Change and the New Geography of Violence* (New York: Nation, 2011).
11. See the discussion in chapter 1, "Messianic White Man." Also see J. Kameron Carter, "Between W. E. B. Du Bois and Karl Barth: The Problem of Modern Political Theology," in *Race and Political Theology*, ed. Vincent W. Lloyd (Stanford, CA: Stanford University Press, 2012).
12. Roberson uses a double negative, a common feature of Black vernacular English, to characterize his intention of uncovering what European (or White) culture's definitions of nature have covered over (or "smothered"), as well as the cover-ups that have routinely taken place since the early nineteenth century, particularly in regard to the relationship between America's understanding of nature as a bottomless cornucopia that can be exploited, and the history and ongoing bequest of the "Triangle Trade."

John Yau, "The Earth Before the End of the World: Ed Roberson's Radical Departure from Romantic Tradition," Poetry Foundation, 13 July 2011, https://www.poetryfoundation.org/articles/69719/the-earth-before-the-end-of-the-world.

13. Thanks to Susan Pensak for pointing out the correlation of whiteout and takeout.
14. Laurel Kearns, "Christian Environmentalism and Its Opponents in the United States," in *Religion in Environmental and Climate Change: Suffering, Values, Lifestyles*, ed. Dieter Gerten and Sigurd Bergmann (New York: Continuum, 2012), 132–51. See also Kearns, "Cooking the Truth: Faith, the Market, and the Science of Global Warming," in *Eco-Spirit: Religions and Philosophies for the Earth*, ed. Laurel Kearns and Catherine Keller (New York: Fordham University Press, 2007).
15. See Catherine Keller, *Face of the Deep: A Theology of Becoming* (New York: Routledge, 2003), esp. chap. 2, "'Floods of Truth': Sex, Love, and Loathing of the Deep."
16. For a recent theopolitical reconsideration of this humanly "dominion," see the ecological consciousness of the papal encyclical *"Laudato Si': On Care For our Common Home,"* originally delivered in Rome, 24 May 2015, http://w2.vatican.va/content/francesco/en/encyclicals/documents/papa-francesco_20150524_enciclica-laudato-si.html.
17. For an important new materialist/affective analysis of the animacy of things, see Mel Y. Chen, *Animacies: Biopolitics, Racial Mattering, and Queer Affect* (Durham, NC: Duke University Press, 2012).
18. Animals have been coming into their own long before their more recent appearance in ecoreligion and theology, starting especially with the work of Carol J. Adams.
19. "Everything I'll say will consist, certainly not in effacing the limit, but in multiplying its figures, in complicating, thickening, delinearizing, folding, and dividing the line precisely by making it increase and multiply." Jacques Derrida, *The Animal That Therefore I Am*, trans. David Wills (New York: Fordham University Press, 2008), 29.
20. For transdisciplinary theological reflection on this manner of animal, bodily, multiplicitous, and ecological differentiation, if not precisely ecological shift, in Derrida's thinking, see one of the latest volumes from the Drew Transdisciplinary Theological Colloquium series, *Divinanimality: Animal Theory, Creaturely Theology*, ed. Stephen D. Moore and Catherine Keller (New York: Fordham University Press, 2014).
21. Jacques Derrida, *The Beast and the Sovereign*, vol. 1, trans. Geoffrey Bennington (Chicago: University of Chicago Press, 2009), 65.

22. Brian Massumi, *What Animals Teach Us About Politics* (Durham, NC: Duke University Press, 2014), 89.
23. Donovan O. Schaefer, *Religious Affects: Animality, Evolution, and Power* (Durham, NC: Duke University Press, 2015).
24. Roberson, "To See the Earth Before the End of the World," 3.
25. Elizabeth Kolbert, *The Sixth Extinction: An Unnatural History* (New York: Henry Holt, 2014), 268.
26. See Paul Crutzen and Eugene F. Stoermer, "The 'Anthropocene,'" *IGBP Newsletter* 41 (May 2000): 17–18; see also Crutzen, "Geology of Mankind: The Anthropocene," *Nature* 415 (2002): 23; and Christophe Bonneuil and Jean-Baptiste Fressoz, *The Shock of the Anthropocene: The Earth, History, and Us* (New York: Verso, 2016). For a nuanced discussion of the metaphysical and theological implications of the Anthropocene, see Austin J. Roberts "Toward an Earthbound Theology," *Open Theology*, 29 December 2017, https://doi.org/10.1515/opth-2018-0006.
27. For a reflection on the growing influence of the language of the Anthropocene, see Elizabeth Kolbert, "The Anthropocene Debate: Marking Humanity's Impact," *YaleEnvironment360*, 17 May 2010, http://e360.yale.edu/features/the_anthropocene_debate_marking_humanitys_impact.
28. Paul Crutzen, who launched the term, does envision our geoengineering the climate as needed ("Planet B"). See David Appell, "The Ethics of Geoengineering," *Yale Climate Connections*, 13 December 2012, https://www.yaleclimateconnections.org/2012/12/the-ethics-of-geoengineering/.
29. Timothy Morton, *Hyperobjects: Philosophy and Ecology After the End of the World* (Minneapolis: University of Minnesota Press, 2013), 24.
30. Morton, *Hyperobjects*, 7.
31. Morton, *Hyperobjects*, 7.
32. Parenti, *Tropic of Chaos*, 167. This link between territory and value challenges not just capitalism but influential progressive views such as that of the political theorists Hardt and Negri, "whose conception of sovereignty is all politics and no place" (172). In this, Parenti finds them mimicking what Marx called "the annihilation of space by time."
33. With the economist Herman Daly, John Cobb has analyzed externalities as a great instance of the effect of what Whitehead has called

the "fallacy of misplaced concreteness." See Cobb and Daly, *For the Common Good: Redirecting the Economy Toward Community, the Environment, and a Sustainable Future*, 2d ed. (Boston: Beacon, 1994 [1989]).

34. Kathryn Tanner, *Christianity and the New Spirit of Capitalism* (New Haven: Yale University Press, 2018). Originally delivered as the Gifford Lectures on Natural Theology, May 2016, https://www.gifford lectures.org/lectures/christianity-and-new-spirit-capitalism. See especially part 4, "Nothing but the Present," which discusses "rapid-response short-termism."

35. Michael S. Northcott, *Place, Ecology and the Sacred: The Moral Geography of Sustainable Communities* (New York: Bloomsbury, 2015). Northcott's *A Political Theology of Climate Change* (2013) makes, as does Tanner, effective use of Christian priorities to prepare for the collapse of modern time.

36. Tanner, *Christianity and the New Spirit of Capitalism*, esp. part 1, "The New Spirit of Capitalism and a Christian Response."

37. Naomi Klein, *This Changes Everything: Capitalism vs. the Climate* (New York: Simon and Schuster, 2014), 21.

38. Klein, *This Changes Everything*, 21 (my emphasis).

39. Klein, *This Changes Everything*, 21.

40. Yet to give Schmitt his due, he was an early critic of "American-style" capitalism, warning of the supercession of politics by economics. See Northcott's analysis of "The Nomos of the Earth and Governing the Anthropocene," in *A Political Theology of Climate Change*, 201–67.

41. Klein, *This Changes Everything*, 10.

42. Klein, *This Changes Everything*, 10.

43. Klein, *This Changes Everything*, 10.

44. For a transdisciplinary theological reflection on the political possibilities of this planetary public, see a recent volume from the TTC series, *Common Good/s: Economy, Ecology, and Political Theology*, ed. Melanie Johnson-DeBaufre, Catherine Keller, and Elias Ortega-Aponte (New York: Fordham University Press, 2015).

45. Donna J. Haraway, *Staying with the Trouble: Making Kin in the Chthulucene* (Durham, NC: Duke University Press, 2016), 55.

46. Haraway, *Staying with the Trouble*, 55.

47. On Lynn Margulis' "symbiosis" in relation to the Gaia hypothesis, see Catherine Keller, *Cloud of the Impossible: Negative Theology and Planetary Entanglement* (New York: Columbia University Press, 2014), 175. On complexity biologist Stuart Kauffman's earlier "autopoiesis," with its Kantian origin, see Keller, *Face of the Deep*, 189ff.
48. Haraway, *Staying with the Trouble*, 55.
49. Haraway, *Staying with the Trouble*, 55.
50. Shelly Rambo, *Resurrecting Wounds: Living in the Afterlife of Trauma* (Waco, TX: Baylor University Press, 2017).
51. Rambo, *Resurrecting Wounds*, 2.
52. Rambo, *Resurrecting Wounds*, 153.
53. Haraway, *Staying with the Trouble*, 1.
54. Haraway, *Staying with the Trouble*, 55–56.
55. See my late twentieth-century reflection *Apocalypse Now and Then: A Feminist Guide to the End of the World* (Boston: Beacon, 1996). At present, I am working on a project called "Apocalypse After All?"
56. Haraway, *Staying with the Trouble*, 4.
57. T. S. Eliot, in "East Coker," in *Four Quartets* (New York: Harcourt, 1971 [1943]), 23–34.
58. Haraway, *Staying with the Trouble*, 137.
59. Thomas Berry, cited in Eileen Crist, "On the Poverty of Our Nomenclature," in *Anthropocene or Capitalocene?: Nature, History, and the Crisis of Capitalism*, ed. Jason W. Moore (Oakland, CA: PM, 2016), 27. See also *Thomas Berry: Selected Writings on the Earth Community*, ed. Mary Evelyn Tucker and John Grim (Maryknoll, NY: Orbis, 2014).
60. See John Grim and Mary Evelyn Tucker, *Ecology and Religion* (Washington, DC: Island, 2014).
61. Crist, "On the Poverty of Our Nomenclature," 27.
62. Berry, in Crist, "On the Poverty of Our Nomenclature," 27.
63. Jennifer Maidrand, the principal organizer of the community garden and a student at Drew Theological School, writes,

> In the past, some food has gone to local food banks, but more recently to feed students on campus (where we have found there is food insecurity). . . . Students enjoy learning agricultural skills, and many enjoy just simply putting their hands in the soil and

interacting with it. The experience of witnessing (and participating in) food growing has been powerful for many students (and children) involved. . . . We are also making small efforts to introduce heritage, biblical, and regional/beneficial crops for remembrance of our past/traditions, and seeking to nourish land we currently inhabit.

Personal correspondence, February 2018.
64. Rambo, *Resurrecting Wounds*, 4.
65. Cusa is citing, with approval, Plato's *Timaeus*. Nicholas of Cusa, "On Learned Ignorance," in *Selected Spiritual Writings*, trans. H. Lawrence Bond (Mahwah, NJ: Paulist Press, 1997), 162.
66. Cusa, "On the Vision of God," in *Selected Spiritual Writings*, 246; 244.
67. See Mary-Jane Rubenstein's multiversal meditations on the holographics of Cusanic cosmology in *Worlds Without End: The Many Lives of the Multiverse* (New York: Columbia University Press, 2014).
68. Cusa, "On Learned Ignorance."
69. Jacques Derrida, *Specters of Marx: The State of the Debt, the Work of Mourning, and the New International*, trans. Peggy Kamuf (New York: Routledge, 2006 [1994]), 45.
70. 1 Cor. 12:27. The "Christ" of this "body of Christ," like all uses of *Christ* in the Bible, is just the Greek translation of Messiah.
71. Jacob Taubes, *The Political Theology of Paul*, trans. Dana Hollander (Stanford, CA: Stanford University Press, 2003), 73. Taubes, who had spent much time in Israel, continues: "He doesn't write: Dear Friend, Nice weather here, or: Glorious nature all around me—he doesn't notice any of that. . . . And yet nature is a very important category—an eschatological category. It groans."
72. See, for example, David G. Horrell, Cherryl Hunt, and Christopher Southgate, *Greening Paul: Rereading the Apostle in a Tme of Ecological Crisis* (Waco, TX: Baylor University Press, 2010).
73. Brigitte Kahl, *Galatians Re-Imagined: Reading with the Eyes of the Vanquished* (Minneapolis: Fortress, 2010), 272.
74. See Haraway, *Staying with the Trouble*, 137, 216n4.
75. On Paul vs. marriage and reproductive family life, see the very passage this book dwells on, 1 Corinthians 7. There is a rich tradition of

feminist critique of Paul's misogyny and homophobia, but also of its contextualization and often false superimposition. The early classic in the latter genre is Elisabeth Schüssler Fiorenza, *In Memory of Her: A Feminist Theological Reconstruction of Christian Origins* (New York: Crossroad, 1994 [1984]). If much of the epistles' worst sexism seems to be based on later textual intercalations, the problem does not therefore ever disappear. Neither, however, does Paul's nonreproductive and egalitarian radicality. See Joseph A. Marchal, ed., *Studying Paul's Letters: Contemporary Perspectivs and Methods*, esp. Melanie Johnson-DeBaufre, "Historical Approaches: Which Past? Whose Past?" And Marchal, "Queer Approaches: Improper Relations with Pauline Letters" (Minneapolis: Fortress, 2012).

76. On Arendt's use of the metaphor of natality, see the feminist philosopher of religion Grace Jantzen's *Becoming Divine: Towards a Feminist Philosophy of Religion* (Bloomington: Indiana University Press, 1999), "In Order to Begin: Death and Natality in the Western Imaginary."

77. See Hannah Arendt, "The Seeds of a Fascist International," in *Essays in Understanding (1930–1954): Formation, Exile, and Totalitarianism* (New York: Schocken, 2005).

78. See Gaye Demiryol, "Arendt and Benjamin: Tradition, Progress and Break with the Past," *Journal of the Philosophy of History* (2016): http://booksandjournals.brillonline.com/content/journals/10.1163/18722636 -12341336#.

79. I thank Susan Pensak for making this connection. See my *Face of the Deep* for connections of a tehomic theology to the Lurianic creation (17, 18, 178, 234).

80. For a queer Pauline reading, see Marchal, *Studying Paul's Letters*, in particular his "Queer Approaches."

81. John Wesley, as cited in Catherine Keller, "The Body of Panentheism," in *Panentheism Actoss the World's Taditions*, ed. Loriliai Biernacki and Philip Clayton (Oxford: Oxford University Press, 2014), 75.

82. See especially Charles Hartshorne, *Omnipotence and Other Theologial Mistakes* (Albany, NY: SUNY Press, 1984); and see also its importance in Sallie McFague, *The Body of God: An Ecological Theology* (Minneapolis: Augsburg Fortress, 1993).

83. For an important intervention into philosophical and theological panic over pantheism, see Mary-Jane Rubenstein's *Pantheologies: Gods, Worlds, Monstrosities* (New York: Columbia University Press, 2018).
84. See Catherine Keller, *Intercarnations: Exercises in Theological Possibility* (New York: Fordham University Press, 2017).
85. See Keller, *Cloud of the Impossible*, chap. 4, "Spooky Entanglements: The Physics of Nonseparability."
86. Karen Barad, "What Flashes Up," in *Entangled Worlds: Religion, Science and the New Materialism*, ed. Mary-Jane Rubenstein and Catherine Keller (New York: Fordham University Press, 2017), 60.
87. See Karen Barad, "Posthumanist Performativity: Toward an Understanding of How Matter Comes to Matter," *Signs: Journal of Women in Culture and Society* 28, no. 3 (2003): 801–31. "Particular possibilities for acting exist at every moment, and these changing possibilities entail a responsibility to intervene in the world's becoming, to contest and rework what matters and what is excluded from mattering" (827).
88. Barad, "What Flashes Up," 63.
89. Barad, "What Flashes Up," 73.
90. Barad, "What Flashes Up," 75.
91. Barad, "What Flashes Up," 75.
92. Barad, "What Flashes Up," 75.
93. Walter Benjamin, "Theses on the Philosophy of History," in *Illuminations: Essays and Reflections*, ed. Hannah Arendt, trans. Harry Zohn (New York: Schocken, 2007 [1969]), 263.
94. See Karen Barad, "Nature's Queer Performativity," *Qui Parle* 19, no. 2 (2011): 121–58.
95. Ed Roberson, "We look at the world to see the earth," *To See the Earth Before the End of the World*, 22 (emphasis in the original).

3. THEOLOGY

1. Clayton Crockett, *Radical Political Theology: Religion and Politics After Liberalism* (New York: Columbia University Press, 2011), 143.
2. Crockett, *Radical Political Theology*, 13.
3. "America's Changing Religious Landscape," Religion and Public Life, The Pew Research Center, 12 May 2015, http://www.pewforum.org/2015/05/12/americas-changing-religious-landscape/.

3. THEOLOGY ❧ 203

4. I thank Rick Benjamins for the invitation, and for the book it occasioned, Benjamins, *Catherine Keller's Constructieve Theologie* (Middleburg: Skandalong, 2017).
5. See chapter 2's discussion of Shelly Rambo's theology of "remaining-with" in the wake of trauma. And for a political theology deconstructive of theological and capitalist normativities of success, see Karen Bray, *Grave Attending: A Political Theology for the Unredeemed* (New York: Fordham Univeristy Press, forthcoming). The Drew dissertation was called "Unredeemed: A Political Theology of Affect, Time, and Worth." 2016.
6. For an excellent summation of theodicy as the strained attempt to justify God as good and omnipotent in the face of unmerited suffering, see David R. Griffin, *God, Power, and Evil: A Process Theodicy* (Philadelphia: Westiminster, 1976).
7. As to the matter of "constructive theology," see, for example, the work of Laurel C. Schneider, in particular in her edited volume with Stephen G. Ray Jr., *Awake to the Moment: An Introduction to Theology, the Workgroup on Constructive Theology* (Louisville, KY: Westminster John Knox, 2016).
8. Crockett, *Radical Political Theology*, 13.
9. Quoted in Peter Baker and Choe Sang-Hun, "Trump Threatens 'Fire and Fury' Against North Korea If It Endangers U.S.," *New York Times*, 8 August 2017.
10. I wrote *Apocalypse Now and Then* mindful of the looming ecological crisis, but even more of Reagan and his friend Hal Lindsay expecting nuclear war in their lifetimes. Catherine Keller, *Apocalypse Now and Then: A Feminist Guide to the End of the World* (Boston: Beacon, 1996).
11. Pankaj Mishra, "The Divided States: Trump's Inauguration and How Democracy Has Failed," *Guardian*, 13 January 2017.
12. Mishra, "The Divided States" (my emphasis).
13. See chapter 1 for my reconstruction of this exceptionalist legacy, as it is analyzed in Kelly Brown Douglas, *Stand Your Ground: Black Bodies and the Justice of God* (Maryknoll, NY: Orbis, 2015).
14. James Baldwin, *The Fire Next Time* (New York: Vintage, 1993 [1962]), 57.
15. Jacques Derrida, *The Gift of Death*, trans. David Wills (Chicago: University of Chicago, 1997). Derrida reflects herein on the politics of the

gift, and particularly "the gift of death," as sacrifice, concerning "the very essence or future of European politics" (33).

16. For a brief interrogation of the gift discourse, see my essay with Stephen D. Moore, "Derridapocalypse," in *Intercarnations: Exercises in Theological Possibility* (New York: Fordham University Press, 2017).
17. Lauren Berlant, *Cruel Optimism* (Durham, NC: Duke University Press, 2012), 24.
18. Gilles Deleuze and Félix Guattari, *What Is Philosophy?*, trans. Hugh Tomlinson and Graham Burchell (New York: Columbia University Press, 1994), 101.
19. Deleuze thus early rendered Kierkegaard's meditation on difference, with the help of Nietzsche: "Repetition is the habitation of difference." See Gilles Deleuze, *Difference and Repetition*, trans. Paul Patton (New York: Columbia University Press, 1995). "Nor should we believe that the contraction is external to what it contracts, or that this difference is external to the repetiton" (286). See also my essay on Deleuze and Whitehead in *Cloud of the Impossible: Negative Theology and Planetary Entanglement* (New York: Columbia University Press, 2014), chap. 5, "The Fold in Process."
20. "It is precisely in the word of the new, the irreducible, that Moltmann finds the end of things in the beginning. Or rather he finds their future, their hoped-for fulfillment, in the 'prereflection of their own true future' (*God and Creation*, 63), the *regnum gloriae*. The phrase *novo creatio ex nihilo* occurs first in *Theology of Hope*." Catherine Keller, *Face of the Deep: A Theology of Becoming* (New York: Routledge, 2003), 245n71.
21. Keller, *Face of the Deep*, 238.
22. Cf. Griffin, *God, Power, and Evil*, in particular his chapter on Calvin, "Calvin: Omnipotence Without Obfuscation."
23. Rabbi Bradley Shavit Artson attends to this crisis with lucid intensity, motivated by his son's autism. See *God of Becoming and Relationship: The Dynamic Nature of Process Theology* (Nashville: Jewish Lights, 2016).
24. Jack (Judith) Halberstam, *The Queer Art of Failure* (Durham, NC: Duke University Press, 2011), 120–21.
25. Halberstam, *The Queer Art of Failure*, 186–87.

26. An Yountae, *The Decolonial Abyss: Mysticism and Cosmopolitics* (New York: Fordham University Press, 2017), 143.
27. An, *The Decolonial Abyss*, 24.
28. From Eckhart's Sermon 83, in *Meister Eckhart: The Essential Sermons, Commentaries, Treatises, and Defense*, trans. Edmund Colledge and Bernard McGinn (Mahwah, NJ: Paulist Press, 1981), 208.
29. Therefore the theology of negation is so necessary to the theology of affirmation that without it God would not be worshiped as the infinite God but as creature; and such worship is idolatry, for it gives to an image that which belongs only to truth itself. God is neither one nor more than one. According to the theology of negation, nothing other than infinity is found in God. Consequently, negative theology holds that God is unknowable either in this world or in the world to come, for in this respect every creature is darkness.

    Nicholas of Cusa, *De Docta Ignorantia*, in *Selected Spiritual Writings*, trans. H. Lawrence Bond (Mahwah, NJ: Paulist Press, 1997), 140 (1.26.86, 88). See also chap. 3 of my *Cloud of the Impossible*, "Enfolding and Unfolding God: Cusanic *Complicatio*."
30. The title of Eckhart's Sermon 13 in *Meister Eckhart: A Modern Translation*, trans. Raymond B. Blakney (New York: Harper and Row, 1957 [1941]).
31. Halberstam, *The Queer Art of Failure*, 24.
32. Alfred North Whitehead, *Adventures of Ideas* (New York: Free Press, 1967 [1933]), 168.
33. Alfred North Whitehead, *Process and Reality: An Essay in Cosmology*, corrected ed., ed. David Ray Griffin and Donald W. Sherburne (New York: Free Press, 1985 [1927–8]), 4.
34. Meister Eckhart, Sermon 6: "The Father gives birth to his Son without ceasing; and I say more: He gives me birth, me, his Son and the same Son." Meister Eckhart, *Meister Eckhart: The Essential Sermons*, 187.
35. "The Adventure of the Universe starts with the dream and reaps tragic Beauty." Whitehead, *Adventures of Ideas*, 296.
36. Mt. 7:21–23.

37. For introductions to the problems of divine omnipotence, passion, and com/passion, see Catherine Keller, *On the Mystery: Discerning Divinity in Process* (Minneapolis: Fortress Press, 2008), chaps. 4–6: "After Omnipotence: Power as Process," "Risk the Adventure: Passion in Process," and "Sticky Justice: Com/passion in Process."
38. The process philosopher Charles Hartshorne sharpened his teacher Whitehead's critique of the classical theist dispassion; see esp. *Omnipotence and Other Theological Mistakes* (Albany: SUNY Press, 1984).
39. For a recent theological meditation on this Einsteinian science of interconnection, see my chapter "Spooky Entanglements: The Physics of Nonseparability," in *Cloud of the Impossible: Negative Theology and Planetary Entanglement* (New York: Columbia University Press, 2014).
40. Alfred North Whitehead, *Science and the Modern World* (New York: Free Press, 1967 [1925]), 174.
41. Whitehead, *Science and the Modern World*, 174.
42. See John B. Cobb Jr., *Christ in a Pluralistic Age* (Eugene, OR: Wipf and Stock, 1998).
43. John D. Caputo, *The Prayers and Tears of Jacques Derrida: Religion Without Religion* (Bloomington: Indiana University Press, 1997). See Clayton Crockett's excellent elucidation of Caputo's radical theology in his chapter 6 on "Caputo's Derridean Gospel" in *Derrida After the End of Writing: Political Theology and New Materialism* (New York: Fordham University Press, 2017), 53–108.
44. John D. Caputo, *The Weakness of God* (Bloomington: Indiana University Press, 2006), 103.
45. See Richard Kearney, *The God Who May Be: A Hermeneutics of Religion* (Bloomington: University of Indiana Press, 2001); and also his edited volume with Matthew Clemente, *The Art of Anatheism* (New York: Rowman and Littlefield, 2018).
46. John D. Caputo, *The Insistence of God: A Theology of Perhaps* (Bloomington: Indiana University Press, 2013), 27. See also Caputo, "If There Is Such a Thing: *Posse ipsum*, the Impossible, and *le peut-être même*. Reading Catherine Keller's *Cloud of the Impossible*," *Journal of Cultural and Religious Studies* 17, no. 1 (December 2017), http://www.jcrt.org/archives/17.1/Caputo.pdf.

47. Cf. Catherine Keller, "The Becoming of Theopoetics: A Brief, Incongruent History," in *Intercarnations*, esp. pp. 111–14.
48. *St. Paul Among the Philosophers*, ed. John D. Caputo and Linda Alcoff (Bloomington: University of Indiana Press, 2009).
49. James H. Cone, *The Cross and the Lynching Tree* (Maryknoll, NY: Orbis, 2011).
50. Robert Kolb, "Luther on the Theology of the Cross," in *The Pastoral Luther: Essays on Martin Luther's Practical Theology*, ed. Timothy Wengert (Grand Rapids, MI: Eerdmans, 2009), 41.
51. "The important thing about this for Luther is that God's living incarnation, sacramentally indwelling "in, with, and under" creation (as Luther will repeat time and time again) is a complex interplay of God revealed and hidden. So, while God's indwelling in the suffering of creation says something about who God is, it also reveals it as beyond human grasp as well (backside of God-like-glory)." Jake Erickson, personal email, March 2018.
52. Jacob Taubes, "Seminar Notes on Walter Benjamin's 'Theses on the Philosophy of History,'" in *Walter Benjamin and Theology*, ed. Colby Dickinson and Stéphane Symons (New York: Fordham University Press, 2016), 184.
53. Taubes, "Seminar Notes," 184.
54. Hille Haker, "Walter Benjamin and Christian Critical Ethics—a Comment," in *Walter Benjamin and Theology*, 299.
55. Judith Butler, "One Time Traverses Another: Benjamin's 'Theological-Political Fragment,'" in *Walter Benjamin and Theology*, 278.
56. Walter Benjamin, "Theologico-Political Fragment," in *Reflections: Essays, Aphorisms, Autobiographical Writings*, ed. Peter Demetz (New York: Schocken, 2007 [1978]), 313.
57. Butler, "One Time Traverses Another," 279.
58. Like a musical note, each bit of matter is "the outcome of vibrations" within "an organized system of vibratory streaming of energy." Whitehead, *Science and the Modern World*, 35.
59. Butler, "One Time Traverses Another," 279.
60. Walter Benjamin, "Theses on the Philosophy of History," in *Illuminations: Essays and Reflections*, ed. Hannah Arendt, trans. Harry Zohn (New York: Schocken, 2007 [1969]), 263.

61. Sharon V. Betcher, "Crip/tography: Disability Theology in the Ruins of God," *Journal for Cultural and Religious Theory* 15, no. 2 (2016): 113.
62. "Earthbodies" is first of all the phrase of the phenomenologist Glen A. Mazis; see his *Earthbodies: Rediscovering Our Planetary Senses* (Albany, NY: SUNY Press, 2002).
63. Betcher, "Crip/tography," 114; Neil Marcus, "The Metaphor of Wind in Cripple Poetics," in *Cripple Poetics: A Love Story*, ed. Petra Kuppers and Neil Marcus (Ipsilanti, MI: Homofactus, 2008).
64. See Augustine's *Confessions*, XII. *The Confessions of Saint Augustine*, trans. John K. Ryan (Garden City, NY: Image, 1960), 358. For commentary on Augustine's hermeneutical pluralism with respect to the creaturely multiplicity of Genesis, see also my *Face of the Deep*, esp. chaps. 2 and 4, "'Floods of Truth': sex, love, and loathing of the deep" and "'Mother most dear': Augustine's dark secrets."
65. Whitehead, *Process and Reality*, 346.
66. Cf. Keller, *On the Mystery*, esp. chaps. 4–6.
67. Whitehead, *Process and Reality*, 342.
68. Whitehead, *Process and Reality*, 342.
69. Whitehead, *Process and Reality*, 342.
70. Whitehead, *Process and Reality*, 343.
71. For a process theology in respectful exchange with more orthodox traditions on the matter of omnipotence, see Thomas Oord, *The Uncontrolling Love of God: An Open and Relational Account of Providence* (Downers Grove, IL: Intervarsity, 2015).
72. Whitehead, *Process and Reality*, 36.
73. Gilles Deleuze, *The Fold: Leibniz and the Baroque*, trans. Tom Conley (Minneapolis: University of Minnesota Press, 1993 [1988]), 81.
74. Whitehead, *Process and Reality*, 50.
75. Whitehead, *Process and Reality*, 31.
76. Whitehead, *Process and Reality*, 94.
77. Eckhart, *Meister Eckhart: The Essential Sermons*, 187.
78. Whitehead, *Process and Reality*, 222.
79. For in-depth reflection on the relation between the impersonal matrix, ground or abyss of mystical traditions Christian or Asian, and Whitehead's theism, see the writings of Joseph A. Bracken, *The Divine Matrix: Creativity as Link Between East and West* (Delhi: Motilal Banarsidass, 1997).

80. Roland Faber and I have in different ways insisted upon the apophatic/mystical supplement to the constructive work of Whiteheadian theology. See especially Faber, *The Divine Manifold* (Lanham, MD: Lexington, 2014); and *God as Poet of the World: Exploring Process Theologies* (Louisville, KY: Westminster John Knox, 2008).
81. "Then the LORD said to Moses, "I am going to come to you in a dense [dark] cloud in order that the people may hear when I speak with you and so trust you" (Ex 19:9). See Keller, *Cloud of the Impossible*, esp. chap. 2, "Cloud-Writing: A Genealogy of the Luminous Dark."
82. See Grace Ji-Sun Kim and Hilda P. Koster, eds., *Planetary Solidarity: Global Women's Voices on Christian Doctrine and Climate Justice* (Minneapolis: Fortress, 2017).
83. See the conclusion to this volume, "Apophatic Afterword," on the process effects of a California-based movement in China.
84. Whitehead, *Process and Reality*, 348.
85. See Mary-Jane V. Rubenstein, *Pantheologies: Gods, Worlds, Monstrosities* (New York: Columbia University Press, 2018).
86. Rick Benjamins, "Apophatic Panentheism: Catherine Keller's Constructive Theology," *Neue Zeitschrift für systematische Theologie und Religionsphilosophie* 60, no. 1 (2018): 103–21, https://doi.org/10.1515/nzsth-2018-0006.
87. Nicholas of Cusa, "On Learned Ignorance," in *Selected Spiritual Writings*, 140 (II.5.118).
88. Wenck cited in Mary-Jane V. Rubenstein, *Worlds Without End: The Many Lives of the Multiverse* (New York: Columbia University Press, 2014), 86.
89. "In [God] we live and move and have our being" (Acts 17:28).
90. Cusa links Paul's apophatic "unknown God" (with reference to the Acts original and the Pseudo/nym Dionysius) to the contrasting Pauline panentheism: "Nevertheless, Paul says, God is not far from anyone, for in God we exist, live and move." "On Seeking God," in Nicholas of Cusa, *Selected Spiritual Writings*, 217.
91. I thank Wendy Lochner for this astute formulation.
92. Giorgio Agamben, *The Time That Remains: A Commentary on the Letter to the Romans*, trans. Patricia Dailey (Stanford, CA: Stanford University Press, 2005), 75.

93. Agamben, *The Time That Remains*, 75. See my discussion of this dynamic of *recapitulatio* in both *Face of the Deep* (esp. pp. 55–56, 221) and *Cloud of the Impossible* (esp. pp. 301–2).
94. *Irenaeus Against Irenaeus*, cited in Keller, *Face of the Deep*, 55ff.
95. Agamben, *The Time That Remains*, 75.
96. See John B. Cobb Jr., *Christ in a Pluralistic Age* (Eugene, OR: Wipf and Stock, 1998 [1975]).
97. See John B. Cobb Jr., *Jesus' Abba: The God Who Has Not Failed* (Minneapolis: Fortress, 2015).
98. N. T. Wright, *The Resurrection of the Son of God* (Minneapolis: Fortress, 2003), 232.
99. Jürgen Moltmann, *The Trinity and the Kingdom: The Doctrine of God*, trans. Margaret Kohl (Minneapolis: Fortress, 1993 [1980]), 131.
100. Linn Tonstad, *God and Difference: The Trinity, Sexuality, and the Transformation of Finitude* (New York: Routledge, 2016), 231.
101. See chapter 5, "'I am because we are': The Roots of Multiplicity in Africa," in Laurel C. Schneider, *Beyond Monotheism: A Theology of Multiplicity* (New York: Routledge, 2008), 53–72.
102. Schneider, *Beyond Monotheism*, 238.
103. "Not everyone who says to me, 'Lord, Lord,' will enter the kingdom of heaven, but only the one who does the will of my Father in heaven" (Mt. 7.21).
104. Jacob Taubes, *The Political Theology of Paul*, trans. Dana Hollander (Stanford, CA: Stanford University Press, 2003), 52f.
105. L. L. Welborn, *Paul's Summons to Messianic Life: Political Theology and the Coming Awakening* (New York: Columbia University Press, 2015), 69.
106. Alain Badiou, *Saint Paul: The Foundation of Universalism*, trans. Ray Brassier (Stanford, CA: Stanford University Press, 2003).
107. Welborn, *Paul's Summons to Messianic Life*, 69.

APOPHATIC AFTERWORD

1. Pseudo-Dionysius, *The Mystical Theology*, in *Pseudo-Dionysius: The Complete Works*, trans. Colm Luibheid (Mahwah, NJ: Paulist Press, 1987), 141.

2. I have elsewhere indicated that the racial resonances of the orthodox "light supremacism" lose all innocence in modernity, as "darkness over the face of the deep" could yoke all dark skins with the chaos to be reduced to nothingness. See Catherine Keller, *Face of the Deep: A Theology of Becoming* (New York: Routledge, 2003), esp. chap. 12, "*Docta ignorantia:* darkness on the face."
3. Fredric Jameson, *The Political Unconscious: Narrative as a Socially Symbolic Act* (NY: Cornell University Press, 1981).
4. On the relation between mystical apophasis and Asian spiritualities, see, for example, D. T. Suzuki's classic *Mysticism: Christian and Buddhist* (New York: Routledge, 2002 [1957]).
5. Nicholas of Cusa, *On Interreligious Harmony: Text, Concordance, and Translation of De Pace Fidei*, ed. by James E. Biechler and H. Lawrence Bond (New York: Edwin Mellen, 1990), 6.
6. William E. Connolly, *The Fragility of Things: Self-Organizing Processes, Neoliberal Fantasies, and Democratic Activism* (Durham, NC: Duke University Press, 2013), 9. See also Connolly, *Aspirational Fascism: The Struggle for Multifaceted Democracy Under Trumpism* (Minneapolis: University of Minnesota Press, 2017).
7. See Jeorg Rieger, *Unified We Are a Force: How Faith and Labor Can Overcome America's Inequalities* (St. Louis, MO: Chalice, 2016), esp. chap. 3, "From Advocacy to Deep Solidarity."
8. Gary Dorrien, *Democratic Socialism: Political Theology, Marxism, and Social Democracy* (New Haven: Yale University Press, 2019).
9. See China Miéville, *October: The Story of the Russian Revolution* (New York: Verso, 2017), 305.
10. See *Salvage*, "About Us," http://salvage.zone/about/.
11. Miéville, *October*, 306.
12. Personal comment to a skeptical, twenty-four-year-old version of myself by Jürgen Moltmann, one that has stuck with me ever since.
13. See, for example, Justin Worland, "Planet Earth's 'Doomsday Clock' Lurches Closer to Midnight Thanks to President Trump," in *Time*, 26 January 2018, http://time.com/4650438/doomsday-clock-donald-tump-atomic-scientists/.
14. Frans de Waal makes a cogent case for the sophistication of nonhuman animal intelligence. But even after serious consideration of the

apparent exception of the parrot, he considers humans "the only linguistic species." *Are We Smart Enough to Know How Smart Animals Are?* (New York: Norton, 2016), 106.

15. Diane Toomey, "Exploring How and Why Trees 'Talk' to Each Other," *YaleEnvironment360*, 1 September 2016, https://e360.yale.edu/features/exploring_how_and_why_trees_talk_to_each_other. Also see Daniel Chamovitz, *What a Plant Knows: A Field Guide to the Senses* (New York: Scientific American, 2012).

16. See Anna L. Tsing, *The Mushroom at the End of the World: On the Possibility of Life in Capitalist Ruins* (Princeton: Princeton University Press, 2015). See also her volume with Heather A. Swanson, Elain Gan, and Nils Bubandt, eds., *Arts of Living on a Damaged Planet: Ghosts and Monsters of the Anthropocene* (Minneapolis: University of Minnesota Press, 2017), a call to revitalize curiosity, observation, and transdisciplinary engagement with life on earth. Karen Barad and Donna Haraway, key to much of the planetary thinking of this political theology and especially its second chapter, both have essays in this volume.

17. Donna J. Haraway, *Staying with the Trouble: Making Kin in the Chthulucene* (Durham, NC: Duke University Press, 2016), 130.

18. See Stephen D. Moore, *Untold Tales from the Book of Revelation: Sex and Gender, Empire and Ecology* (Atlanta: SBL, 2014). See also his edited volume with Laurel Kearns for the TTC series, *Divinanimality: Animal Theory, Creaturely Theology* (New York: Fordham University Press, 2014).

19. Moore, *Untold Tales*, 209.

20. Moore, *Untold Tales*, 202.

21. Moore, *Untold Tales*, 214.

22. See Stephen D. Moore, *Gospel Jesuses and Other Nonhumans: Biblical Criticism Post-poststructuralism* (Atlanta: SBL, 2017).

23. I began to link apophatic theology with race issues in chapter 12, "Docta ignorantia: darkness on the face," in *Face of the Deep*.

24. William Barber, alumnus of Drew Theological School, has ministered to the Greenleaf Christian Church (Disciplines of Christ) in Goldsboro, North Carolina, for over two decades. A member of the board of the NAACP, leader of the "Moral Mondays" campaign, founder of "Repairers of the Breach," and leader of "the new Poor People's Campaign," he has become in recent years a leading progressive voice in

American religious and political life: https://www.breachrepairers.org/about#rev-barber.
25. See my meditation on the negative theopoetics of Whitman's writing in Keller, *Cloud of the Impossible: Negative Theology and Planetary Entanglement* (New York: Columbia University Press, 2014), chap. 6, "'Unfolded Out of the Folds': Walt Whitman and the Apophatic Sex of the Earth."
26. J. William Whedbee, *The Bible and the Comic Vision* (Cambridge: Cambridge University Press, 1998)); Bruce Zucker, *Job the Silent: A Study in Historical Counterpoint* (New York: Oxford University Press, 1991). Also my "tehomophilic" exegesis of Job in *Face of the Deep*, chap. 7, "'Recesses of the Deep': Job's comi-cosmic epiphany."
27. Job 38–41. This does not include the closing frame of the rewards to Job, from the apparent folk narrative that occasions the poet's deconstruction of theodicy *avant la lettre*.
28. I thank my friend Tamara Cohn Eskenazi, professor of biblical literature and history, for handing me the gift of *tiqvah*. The word derives from the root *qavah* meaning "to collect." *Tiqvah* signifies literally a "line" or "cord," a collection of fibers that are twisted together to make a strong cord. As "hope," it refers then not to an abstract future but to an entwinement that binds and strengthens. The proof text is Joshua 2:18 and 21, where it refers to the rope or thread that Rahab is to tie to her window. See Tamara Cohn Eskenazi and Tikva Frymer-Kensky, *JPS Bible Commentary Ruth* (Philadelphia: Jewish Publication Society, 2011), 14–15.
29. Kenneth Ngwa describes this hope as the "weaving and knitting of otherwise fragmented pieces (human bodies, environments) into new life forms defined by trauma-hope." The title of Ngwa's book in process is "Let My People Live: Towards an Africana Reading of Exodus."
30. John B. Cobb Jr., *Is It Too Late? A Theology of Ecology* (Denton, TX: Environmental Ethics, 1995 [1971]); Cobb and Charles Birch, *The Liberation of Life: From the Cell to the Community* (Denton, TX: Environmental Ethics, 1990); John B. Cobb Jr. and Herman E. Daly, *For the Common Good: Redirecting the Economy Toward Community, the Environment, and a Sustainable Future*, 2d ed. (Boston: Beacon, 1994 [1989]).
31. Pando Populus: Where big ideas come down to Earth, "About," https://pandopopulus.com/about/pando-the-tree/.

32. John B. Cobb Jr. and Ignacio Castuera, eds., *For Our Common Home: Process-Relational Responses to Laudato Si'* (Claremont, CA: Process Century, 2015).
33. Philip Clayton and Justin Heinzekehr, *Organic Marxism: An Alternative to Capitalism and Ecological Catastrophe* (Claremont, CA: Process Century, 2014).
34. Toward Ecological Civilization, "Our Mission," http://ecociv.org/about/our-mission/.
35. Monica Coleman, *Bipolar Faith: A Black Woman's Journey with Depression and Faith* (Minneapolis: Fortress, 2016).
36. Martin Luther King Jr., "Beyond Vietnam: A Time to Break Silence," address delivered at Riverside Church, New York, 4 April 1967. In James M. Washington, ed., *A Testament of Hope: The Essential Writings and Speeches of Martin Luther King, Jr.* (New York: HarperCollins, 1991 [1989]), 231–44.
37. Thanks to David E. Roy, Ph.D, for those inspired words (informal correspondence) in response to recent ornithography. He was reading Barbara J. King, "Swooping Starlings in Murmuration," *NPR*, 4 January 2017, https://www.npr.org/sections/13.7/2017/01/04/506400719/video-swooping-starlings-in-murmuration.

INDEX

Abelard, Peter, 155
Abraham, 10, 40, 66, 163
Action, 169–71; language and, 174–75
Acts, 135, 138, 145, 209nn89–90
Adam, 145, 147
Adams, Carol J., 196n18
Adorno, Theodor W., 186n14
Aesop, 52–53, 88
Affects, 80
Agamben, Giorgio, 3, 9, 51–53, 60, 130; christology of, 148–49
Agape. See Love
Agonism, 17–18, 27, 186n14; democratic, 19, 26, 54; as struggle, 28, 30; time related to, 32. See also Amorous agonism
Agonistic respect, 26–27, 186n14
All in, 142–48
Amorous agonism, 54–55, 73, 163; contractions and, 139; transcendence and, 115
Amorous awakening, 154–58
Angel, 54–55, 93, 112

Anglo-Saxons, 46–48
Animal intelligence, 211n14
Animals, 166–69; political, 78–80, 87. See also Nonhumans
Anselm of Canterbury, 155
Antagonism, 21, 26, 39; capitalism and, 28; general, 31–32; political as, 23–24; race and, 49; sovereignty and, 59–60
Anthropic exceptionalism, 6, 75
Anthropocene, 72, 81, 197n28; Capitalocene and, 82–84, 197n32; Chthulucene as, 87–89, 91; Ecozoic and, 91–92
Anthropocene crash, 71–72
Anthropocentrism, 78, 172–73
Anthropos, 6
An Yountae, 119–20
Apocalypse, 69, 89, 148, 158, 194n5; exceptionalism and, 109; geopolitics and, 21, 27, 67; triple, 107, 110, 163
Apocalyptic Christian right, 2
Apophasis ("unsaying"), 15

Apophatic Action/Theology, 169–71
Apophatic Animality/Earth, 166–69
Apophatic Assemblage/Political, 163–66
Apophatic darkness, 141
Apophatic ellipsis, 17
"Apophatic entanglement," 15–16
Apophatic panentheism, 132
Arendt, Hannah, 97–98, 286n14
Aristotle, 80, 125–26
Artson, Bradley Shavit, 204n23
Atheism, 117, 145
Augustine, 12, 43, 97, 135, 147, 189n64
Awakening, amorous, 154–58

Babylonian, 75–76
Bacon, Francis, 14
Badiou, Alain, 156
Bakunin, Mikhail, 8–9
Baldwin, James, 110–12
Barad, Karen, 18, 104–6, 140, 212n16; on Benjamin, 102–3, 131; quantum indeterminacy of, 101–2, 134, 202n87
Barber, William, 170–71, 212n24
Beckett, Samuel, 122–23, 133, 160
Becomingness, 42–44
Becomings, 139–40, 147–48
Belief, 125
Benjamin, Walter, 4, 17, 51–52, 132, 150; angel of, 54–55, 93, 112; Barad on, 102–3, 131; enormous abridgement of, 133; Halberstam on, 118; *Jetztzeit* of, 4, 60, 98, 148; kabbalistic mysticism of, 98, 101; on Messiah, 130–31; time and, 60–61, 98
Bennett, Jane, 36
Berlant, Lauren, 114
Berry, Thomas, 91
Betcher, Sharon, 134–35
Bible, 42, 110, 113, 169, 173. *See also specific books*
Birds, 179
Bloch, Ernst, 56–59, 113, 137
Bray, Karen, 106
Brilliant darkness, 120–22
Brown Douglas, Kelly, 46, 68, 111
Buddha, 175–76
Buddhist nonattachment, 90
Bush, George W., 52–53
Butler, Judith, 17, 101, 131

Caesar, 60, 137, 151–52
Calvinism, 116
Capitalism, 19, 26–28, 84, 187n18, 198n40
Capitalist-evangelical resonance machine, 27–28, 83
Capitalist theodicy, 33–35
Capitalocene, 82–84, 197n32
Caputo, John D., 127–29
Carter, J. Kameron, 50
Catastrophe, 2, 14, 54, 91, 110, 112, 173
Catechontic sovereignty, 62
Certainty, 14–15
Chaotic edge of creation, 114–16
Charisma, 66; amorous agonism and, 54–55; international law and, 51–53; juridico-political

system and, 52; secularizations of, 53–54; Weber on, 53
China, 83, 176
Christ, 148–54. *See also* Jesus Christ
Christianity, 2, 7, 157; Caesar and, 137; exceptionalism of, 18, 75, 105; exhaustion of, 106–7; Judaism and, 50; time related to, 84, 198*n*35
Christian political theology, 24–25
Christian right, 2–3, 109, 111
Christian social utopias, 137
Christology, 148–49
Chronos (measured time), 3–4, 149, 181*n*3
Chthulucene, 87–89, 91
Civility, 26
Clayton, Philip, 176–77
Climate change, 5, 18, 35–36, 73, 82, 175
Climate Change in Mountain Ecosystems Project, 71
"Climate fascism," 74
Climate stability, 81
Coalition, 64
Cobb, John, Jr., 11–13, 126, 142, 150, 197*n*33; lateness of, 175, 179; secularism of, 175–78
Cohen, Leonard, 60
Coincidentia oppositorum (coinciding opposites), 66–67
Colbert, Stephen, 80
Coleman, Monica, 178
"Common home," 21–22, 92
Commons, 86; undercommons, 30–32, 64, 72

"Communitarian particularisms," 156
Comparative theology, 10
Concept of the Political, The (Schmitt), 23
Cone, James, 129
Confessions (Augustine), 135
Confucius, 12, 175–76
Connectivity, 155–56, 210*n*103. *See also* Ethos of interconnectedness
Connolly, William, 17, 34–35, 164; agonistic respect of, 26–27, 186*n*14; resentment and, 27, 186*n*14, 187*n*18
Contracted (*sunestalemnos*), 3, 22, 32, 61, 85, 98
Contracted infinity, 93–95
Contraction, practice of, 63–68
Contractions, 159; amorous agonism and, 139; becomings in, 139; differences from, 139–40; of earth, 99–100; of economics, 85–86; insurrectionist theology and, 63–64; integration as, 91; issue/coalition and, 64; kairos and, 2–5; locality/globality and, 64–65; negotiation as, 121–22; of polis, 7; of political, 22; religion/saeculum, 65–66; schema of, 7; of time, 61; of universe, 6, 100, 104, 137–38
Contradictions, 165
Cord (*tiqvah*), 173–74, 178, 213*n*28
Corinthians, 2, 5, 32, 62, 67, 148, 200*n*75; power in, 128–29; time of, 3–4, 61

Cosmic ecology, 95–100
Cosmologies, 123
Counterexceptionalist theology, 177–78; God of the counterexception, 136–42, 154; murmuration and, 179–80
Covenant: of hope, 113–14; of justice, 113
Creation, 42–43, 114; collapse of, 108, 110; dominion over, 75–77, 167; eschatology and, 115–16; in exception, 75–77; God within, 124–25, 127–28, 130, 138, 145, 207n51; good of, 76; Paul on, 96–97
Creativity, 119
Crisis, 39, 204n20, 227; kairos as, 68; "staying with," 88–90, 112, 168; triune, 108–10, 163
Crist, Eileen, 91
Critical animal theory, 78
Critical difference, 39–41
Critical theory, 131
Critters. *See* Nonhumans
Crockett, Clayton, 63–64, 67, 79, 105, 109
Crucifixion, 55–56, 129, 168–69
Cruising Utopia (Muñoz), 56
Crutzen, Paul, 197n28
Cusa. *See* Nicholas of Cusa

Daly, Herman, 197n33
Daly, Mary, 12
Dark cloud, 94, 99, 141, 143–44, 209n81
Darkening hope, 15–17
Darker brilliance, 160–63

Darkness, 120–22, 141, 161, 162, 211n2
Darkness on the face (*tehom*), 34, 188n35
David (king), 113
Death: of God, 107, 109, 117, 124, 127, 130, 133; of noncombatants, 58, 193n98
Debt, 32
Decisions, 31, 39–40, 42–43
Deconstruction, 128, 166
Deification (*theosis*), 144
Deleuze, Gilles, 79, 114, 140, 149, 204n19
Democracy, 140; Messiah and, 79; social, 164–66; Trump and, 111
Democratic agonism, 19, 26, 54
Derrida, Jacques, 17, 58, 78–79, 95, 196n19; Caputo on, 127; on gift, 113, 203n15; Messiah and, 130; metaphysics of presence of, 138
Deterritorialization, revolutions and, 114
Differences, 29, 39–41; in anthropocene crash, 71–72; from contraction, 139–40; divinanimality and, 78–79
Dionysius the Areopagite, 161–62
Disorientation, 122–23
Divinanimality, 78–79, 94, 96, 113, 169
Divine, 76, 119–20. *See also* God
Docta Ignorantia, De (Cusa), 144
Douglas, Kelly Brown, 18
Dualism, 140–41

Earth: Apophatic Animality, 166–69; contraction of, 99–100; humans on, 72, 76–77; interhuman crises on, 72; as matter, 72; poetry of, 69–71, 74, 80–81, 104; as Undercommons, 72
Earth Day (1980), 14
Earth-time, 37
East Africa, 29
Eaton, Heather, 91
Eckhart, Meister, 120, 124, 127–28, 141, 205n34
Ecocene, 160–61
Ecocene inception, 91–93
EcoCiv, 177–78
Ecogeopolitics, 21–22
Economics, 65, 198n40; contraction of, 85–86
Ecosocial justice experiments, 13–14
Ecosocial predictions, 17
Ecotraumas, 108
Ecozoic, 91–92
Edelman, Lee, 56, 97
Ehrlich, Paul, 81
Einstein, Albert, 126
Eliot, T. S., 89, 128
Elohim (God), 42–43, 99, 190n66
Embracing Hopelessness (de la Torre), 89–90
Emergency, 86, 179; Capitalocene as, 82; into emergence, 92; glaciers as, 69–71; inception and, 62–63; solidarity in, 72–73
Empire, 44–46
Empowerment, 129
Enemies, 17–18, 22–26, 55

Energy, 67
Enormous abridgement, 133
Enuma Elish, 75–76
Environment, 73–74
Ephesians, 67, 148
Eros of the universe, 125–28, 136
"Eros of the Universe" (Whitehead), 127–28, 136, 147
Eschatology, 115–16
Ethnicity, 46–48; racialization and, 49–50
Ethos of interconnectedness: antagonism in, 39; aspirational fascism and, 35–37; climate debt in, 38; Connolly on, 35–38; earth-time and, 37; kairos and, 37–38; neoliberalism and, 35–36; pluralist assemblage in, 38; Whitehead and, 36
Eusebius, 45, 153, 190n69
Exception, 50, 72; without, 5–7; Anglo-Saxons and, 46–48; creation in, 75–77; God's counter-, 136–42, 154; Latin and, 48; Schmitt and, 16, 18; sovereign decision of, 31; state of, 51–53; takeout as, 80–81; whitening related to, 46–49. *See also* Sovereign exception
Exceptionalisms, 6, 14, 64, 73–74, 161; of Christianity, 18, 75, 105; moral code and, 109; schematism of, 16–18; secularization of, 112; sexual, 57, 192n95; triune crisis of, 108–10; White, 18, 46–49. *See also* Counterexceptionalist theology

Exodus, 47
Extinction, 69, 91

Faber, Roland, 209n80
Fagre, Daniel, 70, 194n5
Fail better, 122–25, 160
Failing God, 107–9, 124–25, 135–36, 146
Failure, 118–19
False dilemma, 23
Family, 23–24, 186n9
Fascism, 35–37, 41, 102
Feminism, 3, 200n75
Fiorenza, Elisabeth Schüssler, 200n75
Fire Next Time, The (Baldwin), 110
First Urban Christians (Meeks), 7
First World War, 112–13
Foes, 17–18, 22–26
Food, 77, 92–93, 199n63
Form (schema), 4–5, 7, 71, 138–39
For the Common Good (Cobb), 175
Foucault, Michel, 149, 186n14
Fragility of Things, The (Connolly), 34
Francis (Pope), 30, 176; "common home" of, 21–22, 92
Francis (Saint), 53
Franklin, Benjamin, 47–48, 73
Friends and Foes, 17–18, 22–26
Fulfillment, of law, 156–57
Fundamentalism, 66
Futurity, 56

Galatians (5:14), 155
Gandhi, Mohandas, 53, 121–22
Gardens, 92, 199n63

Gathered together (*sunestalemnos*), 3, 22, 32, 61, 85, 98
Geertz, Clifford, 162–63
Gender, 41, 178
General antagonism, 31–32
Genesis, 42–43, 75, 76–77, 135, 167
Geological Survey, U.S., 71
Geopolitics, 21, 27, 67
Glaciers, 69–71, 81, 194n5
Globality, 64–65
God, 96, 171; all in, 145–46; atheism and, 117, 145; belief in, 125; birth of, 124, 205n34; within creation, 124–25, 127–28, 130, 138, 145, 207n51; dark cloud as, 94, 99, 141, 143–44, 209n81; death of, 107, 109, 117, 124, 127, 130, 133; deconstruction of, 128; dualism of, 140–41; Elohim as, 42–43, 99, 190n66; empowerment from, 129; failing, 107–9, 124–25, 135–36, 146; *Hashem* as, 161–62; interconnection and, 126; Kingdom of, 4–5, 62, 182n4, 182n12; laughter of, 172; love of, 120, 158; omnipotence of, 116–17; possibility of, 125–27, 146; as Process, 140; secularization of, 121; Spirit as, 134–35, 156–58; time of, 194n114; weakness of, 128–30, 135–36; as White, 112, 119–20, 138; White void of, 109–12; womb of, 99; world and, 142–44. *See also specific topics*
God of the counterexception, 136–42, 154

God Who May Be (Kearney), 127
Goliath, 150
Grace, 147
Gramsci, Antonio, 286*n*14
Great Transformation, The (Polanyi), 35
Greek, 39, 43, 92, 173, 182*n*4, 200*n*70; *sunestalemnos* and, 3, 22, 32, 61, 85
Gregory of Nyssa, 161–62
Grim, John, 91

Habermas, Jürgen, 286*n*14
Halberstam, Jack, 118, 122
Happiness, 132, 135
Haraway, Donna, 87–91, 97, 105, 212*n*16
Hardt, Michael, 24, 197*n*32
Harney, Stefano, 30
Hartshorne, Charles, 125–26, 206*n*38
Hashem, 161–62
Hayes, Chris, 143
Hazardous Hope, 57–59
Hegel, Georg Wilhelm Friedrich, 40, 130
Heidegger, Martin, 97, 149, 286*n*14
Hillis, Wren, 186*n*14
History, 159; time and, 60–62, 81, 103
Hitler, Adolf, 193*n*98
Holocene, 5, 81, 91, 108, 160–61
Homonationalism, 192*n*95
Homosexuality, 28, 64, 119, 192*n*95, 195*n*9
Hope, 19, 57, 59, 159, 192*n*96, 213*n*29; covenant of, 113–14; darkening of, 15–17; Job and, 173–74, 213*n*28; in ruins, 89–90; secularization of, 58, 85
Hope Draped in Black (Winters), 32–33
Hosea, 85
Humans: connectivity of, 155–56, 210*n*103; divine related to, 76, 119–20; on earth, 72, 76–77; nonhumans and, 5–6, 30, 77–80, 167–69
Humility, 175
"Hunchbacked dwarf," 133–34
Hunchback theology, 131–36

Identity, 25, 156
Immanence, 114–15, 123
Immigration, 73–74
Inception, 43–44; Ecocene, 91–93; emergency and, 62–63; kairos and, 59–63
Inclusivism, 25–26
Infinity, 93–95, 101–2
Insurrectionist theology, 63–64
Integration, 91
Interconnection, 126, 154. See also Ethos of interconnectedness
Interdependence, 161
International law, 51–53
Irenaeus, 149–51
Isaac, 40
Is It Too Late? (Cobb), 175
Islam, 66, 153, 164, 192*n*95
Issue/coalition, 64

James, William, 140
Jerusalem, 21

Jesus Christ, 9, 12, 21, 175–76, 182*n*12; Caesar and, 151–52; connectivity of, 155–56; crucifixion of, 55–56, 129, 168–69; Irenaeus on, 150–51; on light, 47; Second Coming of, 3; subordinationism of, 152–53
Jetztzeit (now-time), 4, 60, 98, 148
Job, 99, 107, 171–72, 213*n*27; *tiqvah* of, 173–74, 178, 213*n*28
John of Patmos, 2
Joshua (2:18 and 2:21), 213*n*28
Judaism, 50, 98, 101, 105–6
Jurecic, Quinta, 52–53
Juridico-political system, 52
Justice, 113

Kafka, Franz, 96
Kahl, Brigitte, 96
Kairos (right time), 17, 182*n*4, 194*n*114; catastrophe and, 2, 14, 54, 91, 110, 112, 173; *chronos* and, 3–4, 181*n*3; contraction and, 2–5; creation and, 43; as crisis, 68; ethos of interconnectedness and, 37–38; fullness of, 18–19; inception and, 59–63; Paul and, 2–3, 22, 28, 42, 182*n*12
Katongole, Emmanuel, 29, 33, 89
Kearney, Richard, 127
Kierkegaard, Søren, 40, 149, 204*n*19
King, Martin Luther, 55, 111, 121–22, 179
Kingdom of God, 4–5, 62, 182*n*4, 182*n*12

Klee, Paul, 54
Klein, Naomi, 85–86, 103, 112
Kolbert, Elizabeth, 81

Lament, 29, 171–72
Language: action and, 174–75; on edge, 122–23; of nonhumans, 167–68, 211*n*14; silence as, 141–42; theology and, 120–21
Laotzu, 12, 175–76
Lateness, 175, 179
Laughter, 172
Law: fulfillment of, 156–57; international, 51–53
Leibniz, Gottfried Wilhelm, 34, 140, 149
Levinas, Emmanuel, 286*n*14
LGBTQ, 28, 64, 119, 195*n*9
Liberalism, 35–36, 41
Light, 47
Lindsay, Hal, 203*n*10
Line (*tiqvah*), 173–74, 178, 213*n*28
Lobeda, Cynthia Moe, 38
Locality/globality, 64–65
Love (*agape*): in Buddhist nonattachment, 90; of enemies, 23–25, 55; of God, 120, 158; of neighbor, 155–57; politics of, 24
Luther, Martin, 130, 207*n*51

Mahler, Gustav, 37
Maidrand, Jennifer, 199*n*63
Malabou, Catherine, 105, 119
Manley, James K., 182*n*13
Margulis, Lynn, 87, 105

Martin, Trayvon, 68
Marx, Karl, 58, 137, 197*n*32
Marxism, 60, 137, 165–66, 176–77, 186*n*14
Massumi, Brian, 79–80, 167–68
Matter, 72, 135; vibrations as, 132, 207*n*58
Matter, Messianism, 100–4
Matthew (25:31–46), 155
McKibben, Bill, 176
Measured time (*chronos*), 3–4, 149, 181*n*3
Meeks, Wayne, 7
Meeting the Universe Halfway (Barad), 101
Messiah, 17, 128–29; Agamben on, 51; all in, 145; Benjamin on, 130–31; democracy and, 79; embodiment of, 67; lamb as, 169; nature and, 132; revolutions and, 56; time and, 61
Messianic redemption, 149
Messianic white man, 49–51
Messianism Matter, 100–4
Metaphysics, 138, 141
Metz, Johannes, 10–11, 131
Micah, 171
Miéville, China, 165–66
Mindfulness, 159
Mishra, Pankaj, 110–11
Mohammed, 12, 175–76
Moltmann, Jürgen, 10–11, 57–58, 153, 204*n*20, 211*n*12
Monarchy, 45
Monotheism, 45, 190*n*69
Monty Python, 121

Moore, Stephen, 168–69
Moral code, 109
Mortal critters entwined, 86–90
Morton, Tim, 37, 82–83
Moses, 12, 175–76, 209*n*81
Moten, Fred, 30–31
Mouffe, Chantal, 17, 23, 26, 54, 186*n*14
Mozart, Wolfgang Amadeus, 132
Muñoz, José Esteban, 56–57, 176
Murmuration, 179–80
Mutual immanence, 123

NAACP, 212*n*24
Nancy, Jean-Luc, 143
Nathan (prophet), 113
Nature, 131–32
Negative political theology, 16, 170
Negative theology, 15, 99, 133, 159, 162, 205*n*29; tradition of, 120–21
Negotiation, 121–22
Negri, Antonio, 24, 197*n*32
Neighbor, love of, 155–57
Neocolonial economics, 65
Neoliberalism, 35–36
Newness, 115, 175, 204*n*20
New Revised Standard Version, 3
New Testament, 3, 155, 182*n*4; Acts, 135, 138, 145, 209*nn*89–90; Philippians, 62; Romans, 156. *See also* Corinthians
Ngwa, Kenneth, 174, 213*n*29

Nicholas of Cusa, 6, 93–95, 120, 126, 205n29; *Peace of the Faith* of, 163–64; theocosm of, 144–45
Nietzsche, Friedrich, 28, 149, 286n14
Nihil aliquid ("Nothingsomething"), 43, 189n64
Nihilism, 2
Noah, 77, 110
Noah, Trevor, 80
Noncombatant death, 58, 193n98
Nonhumans, 88–90; affects related to, 80; birds, 179; crucifixion related to, 168–69; divinanimality of, 78–79, 94, 96, 113, 169; humans and, 5–6, 30, 77–80, 167–69; interdependence with, 161; language of, 167–68, 211n14; political, 78–80, 87, 196n18
Northcott, Michael, 12, 65, 84
Nothingness: Augustine on, 43, 189n64; becomingness or, 42–44; decision from, 42–43; inception within, 43–44
"Nothingsomething" (*nihil aliquid*), 43, 189n64
Novus, Angelus, 62
Now-time (*Jetztzeit*), 4, 60, 98, 148

Omnipotence, 116–17, 161
Optimism, 114
Order (schema), 4–5, 7, 71, 138–39
Organic Marxism, 176–77
Origen, 149
Original sin, 146–47

Other, enemy as, 25
Ottoman Islam, 164

Pale optimism, 112–14
Pally, Marcia, 24
Pando Populus network, 176
Panentheism, 143–45, 209nn89–90
Pantheism, 144
Pantheologies (Rubenstein), 143
Parenti, Christian, 74, 82, 197n32
Parker, Matthew, 46
Parousia, 151–52
Pastoral crisis, 204n20, 227
Paul (apostle), 7, 9, 17; Caputo and, 128–29; on creation, 96–99; crucifixion and, 129; Ephesians and, 67; against feminism, 200n75; household of God of, 96; kairos and, 2–5, 22, 28, 42, 182n12; on *parousia*, 151–52; on *schema*, 71; Schmitt on, 62; *sunestalemnos* of, 3, 85, 98
Paul panentheism, 143, 145, 209nn89–90
Peace of the Faith (Nicholas of Cusa), 163–64
Permafrost, 71
Pew poll, 106
Philippians, 62
Plants, 167–68
Plasticity, 83–85, 105, 142
Pluralism, 13, 38
Poetry, 29, 95; about earth, 69–71, 74, 80–81, 104; theopoetics, 134–35
Polanyi, Karl, 35

Polis, 7, 11, 22, 26, 47
Political: as antagonism, 23–24; Apophatic Assemblage, 163–66; definitions of, 22, 30; false dilemma and, 23; Schmitt on, 22–23; theology within, 10
Political animals, 78–80, 87
Political nonhumans, 78–80, 87, 196n18
Political theology, 7, 11, 13–15; negative, 16, 170; phrase origin of, 8–9; Social Gospel movement and, 9–10, 12
Political Theology (Schmitt), 8–9, 39, 42
"Political Theology of Mazzini and the International, The" (Bakunin), 9
Politics, 27, 63–64, 67; ecogeopolitics, 21–22; of love, 24; religion compared with, 8
Power, 31, 86, 128–31, 198n40
Practice, of contraction, 63–68
Precapitulation, 17–19
Presence, 138
Process, God as, 140
Process and Reality (Whitehead), 137
Process theology, 11, 18, 142, 145, 150, 178; of Whitehead, 138, 175
Process Theology as Political Theology (Cobb), 11
Progressive theology, 13
Promise, 113
Prophetic movements, 12–13
Psalm 8, 152–54

Puar, Jasbir, 57, 192n95
Publications, 176–77

Quantum indeterminacy, 101–2, 134, 202n87
Quantum physics, 139
Queer arts, 117–20, 122
Queer performativity, 56–57

Race, 49, 211n2
Racialization, 23–24, 186n9; *docta ignorantia* and, 34, 188n35; ethnicity and, 49–50
Racism, 47–48, 111; immigration and, 73
Rahab, 213n28
Rambo, Shelly, 88–89, 93
Rashi, 43
Rationalism, 117
Rawls, John, 286n14
Reactionary denialism, 2
Reagan, Ronald, 27, 203n10
Recapitulation, 103, 148–50, 154
Redemption, 149
Religiocultural multiplicity, 163–64
Religion: capitalism and, 27, 187n18; empire related to, 44–46; politics compared with, 8; secularity and, 66, 194n112
Religion/saeculum, 65–66
Religious socialism, 3–4
Repetition, 114, 204n19; recapitulation, 103, 148–50, 154
Resentment, 27–28, 186n14, 187n18
Revelation, Book of, 169

Revolutions: deterritorialization and, 114; Messiah and, 56
Right time. *See* Kairos
Robbins, Jeffrey, 11–12
Roberson, Ed, 18, 74, 95, 111, 133, 195*n*12; takeout of, 80–81; time of, 69–70, 103–4
Romans, 156
Roosevelt, Theodore, 48
Rubenstein, Mary-Jane, 143

Saeculum (time), 8, 12–13
Salvage journal, 165–66
Sanders, Bernie, 165
Satan, 173
Schaefer, Donovan, 80
Schema (form, order), 4–5, 7, 71; of world, 138–39
Schematism, 16–18
Schmitt, Carl, 10–11, 13, 62, 131, 286*n*14; Bakunin and, 8–9; against capitalism, 198*n*40; Christian political theology and, 24–25; exception and, 16, 18; inclusivism and, 25–26; on nothingness, 42; on political, 22–23; on power, 31, 86, 198*n*40; sovereign exception of, 39–41; Trump and, 52–53
Science, 100
Secularism, 8, 12–13, 65–67; of Cobb, 175–78
Secularity, 66, 142, 194*n*112
Secularizations, 142, 159–60; of charisma, 53–54; of exceptionalism, 112; of God, 121; of hope, 58, 85

Self-destruction, 160
Self-organization, 87–88, 157
Self-secularization, 142
Sensuous Intersectionalities, 55, 178; queer performativity in, 56–57
Sexual exceptionalism, 57, 192*n*95
Sexuality, 57; homosexuality, 28, 64, 119, 192*n*95, 195*n*9
Shiva, Vandana, 176
Shock Doctrine (Klein), 86
Short (*sunestalemnos*), 3, 22, 32, 61, 85, 98
Silence, 164; as language, 141–42; science and, 100; unspokenness as, 162
Sin, 146–47
Social democracy, 164–66
Social Gospel movement, 9–10, 12
Socialism, religious, 3–4
Social justice, 12–13
Social movement/electoral politics, 63–64
Social movements, 59–60
Social utopias, Christian, 137
Socrates, 12
Solidarity, 102; in emergency, 72–73
Sölle, Dorothee, 10–11
Solnit, Rebecca, 90
Song of the Earth (Mahler), 37
Sovereign decision, 31
Sovereign exception, 46, 51, 60, 160; decisions and, 39–40; Fascism and, 41; liberalism and, 41; universal in, 40

Sovereignty, 18, 62, 66, 161–62; antagonism and, 59–60; subordinationism and, 152–53; whiteness of, 50
Space, 1, 83–84
Speciesism, 48–49
Spirit, 134–35, 156–58
"Spirit, Spirit of Gentleness" (Manley), 182*n*13
Spivak, Gayatri, 65, 193*n*109
Stalin, Joseph, 193*n*98
Stand Your Ground: Black Bodies and the Justice of God (Brown Douglas), 46
State of Exception (Agamben), 51–52
Stoicism, 4
Struggle, 28, 30
Subordinationism, 152–53
Suffering, 131–32
Sunestalemnos (gathered together, contracted, short), 3, 22, 32, 61, 85, 98
Survival, 29
Symbiogenesis, 87–88
Sympoiesis, 87–88
Synchronicity, 179–80

Tacitus, 46
Takeout, 80–86
Tanner, Kathryn, 40, 83–84, 198*n*35
Taubes, Jacob, 51, 60, 62, 96, 130, 155, 200*n*71
Tehom (chaos, ocean), 43
Theocosm, 144–45
Theodicy, 33–35, 116–20

Theology: Apophatic Action, 169–71; energy as, 67; "hunchbacked dwarf" of, 133–34; language and, 120–21; within political, 10; time for, 105; word of God as, 171. *See also specific types*
Theology of Hope (Moltmann), 57–58
Theopoetics, 134–35
Theopoiesis, 126
Theosis (deification), 144
Thessalonians, 3
This Changes Everything (Klein), 86
Tillich, Paul, 3–4, 12, 182*n*4
Time, 2, 37, 105; agonism related to, 32; Benjamin and, 60–61, 98; Christianity related to, 84, 198*n*35; *chronos* as, 3–4, 149, 181*n*3; contraction of, 61; of Corinthians, 3–4, 61; glaciers and, 70–71, 194*n*5; of God, 194*n*114; history and, 60–62, 81, 103; infinity of, 93–95, 101–2; *Jetztzeit*, 4, 60, 98, 148; plasticity and, 83–85; of Roberson, 69–70, 103–4; *saeculum* as, 8, 12–13; space and, 1, 83–84; synchronicity in, 179–80. *See also* Anthropocene; Kairos
Time-names, 72
Time That Remains, The (Agamben), 3
Tiqvah (cord, hope), 173–74, 178, 213*n*28
Tonstad, Linn, 153–54

de la Torre, Miguel, 89–90
"To See the Earth Before the End of the World" (Roberson), 18
Towne, Emilie, 32
Tradition, of negative theology, 120–21
Transcendence, 143; immanence and, 114–15
Trauma, 29
Triune crisis, 108–10, 163
Trump, Donald, 109–10, 165, 195n9; Democracy and, 111; against environmentalism, 73–74, 167; Schmitt and, 52–53
Trust, 163
Truth, 120
Tsing, Anna, 168
Tucker, Mary Evelyn, 91

Undercommons, 30–32, 64; Earth as, 72
Undercosmos, 101
Universalism, 65
Universe, 125–28, 136, 141; contraction of, 6, 100, 104, 137–38
"Unsaying" (*apophasis*), 15
Unspokenness, 162
Utopia, 58–59

Vibrations, 132, 207n58
Victors, 118
Voltaire, 34

de Waal, Frans, 211n14
Water, 70–71

Weak messianic power, 128–31
Weakness, of God, 128–30, 135–36
Weakness of God, The (Caputo), 127
Weaver's shuttle, 4, 160, 173
Weber, Max, 53
Welborn, L. L., 4, 156, 182n12
Wenck, Johannes, 144
Wesley, John, 99
White, 21, 32, 94, 161; Beckett and, 122–23; Franklin and, 47–48; futurity and, 56; God as, 112, 119–20, 138; messianic white man, 49–51; Roberson and, 195n12
White, Carol Wayne, 48–49
White exceptionalisms, 18, 46–49
Whitehead, Alfred North, 13, 36, 59, 90, 183n16, 197n33; Benjamin and, 132; on Caesar, 137; Cobb and, 126; "Eros of the Universe" of, 127–28, 136, 147; Hartshorn and, 206n38; Leibniz compared to, 140; "mutual immanence" of, 123; "principle of ultimate" of, 141, 209n80; process theology of, 138, 175; recapitulation of, 149–50
Whiteness, 186n9; Anglo-Saxons and, 46–48; of sovereignty, 50; speciesism and, 48–49
Whiteout, 73–74
White supremacists, 57, 97–98, 110–11, 152, 155
White void of God, 109–12
Whitman, Walt, 171

Winters, Joseph, 32–33, 90
World, 112–13, 158; end of, 69;
    God and, 142–43; schema of,
    138–39
Worstward Ho (Beckett), 122–23
Wright, N. T., 151–52

Xi Jinping, 177

Yau, John, 195n12
Yeats, William Butler, 112–13

Žižek, Slavoj, 9

Radical Democracy and Political Theology, Jeffrey W. Robbins

Hegel and the Infinite: Religion, Politics, and Dialectic, edited by Slavoj Žižek, Clayton Crockett, and Creston Davis

What Does a Jew Want? On Binationalism and Other Specters, Udi Aloni

A Radical Philosophy of Saint Paul, Stanislas Breton, edited by Ward Blanton, translated by Joseph N. Ballan

Hermeneutic Communism: From Heidegger to Marx, Gianni Vattimo and Santiago Zabala

Deleuze Beyond Badiou: Ontology, Multiplicity, and Event, Clayton Crockett

Self and Emotional Life: Philosophy, Psychoanalysis, and Neuroscience, Adrian Johnston and Catherine Malabou

The Incident at Antioch: A Tragedy in Three Acts / L'Incident d'Antioche: Tragédie en trois actes, Alain Badiou, translated by Susan Spitzer

Philosophical Temperaments: From Plato to Foucault, Peter Sloterdijk

To Carl Schmitt: Letters and Reflections, Jacob Taubes, translated by Keith Tribe

Encountering Religion: Responsibility and Criticism After Secularism, Tyler Roberts

Spinoza for Our Time: Politics and Postmodernity, Antonio Negri, translated by William McCuaig

Force of God: Political Theology and the Crisis of Liberal Democracy, Carl A. Raschke

Factory of Strategy: Thirty-Three Lessons on Lenin, Antonio Negri, translated by Arianna Bove

Cut of the Real: Subjectivity in Poststructuralism Philosophy, Katerina Kolozova

A Materialism for the Masses: Saint Paul and the Philosophy of Undying Life, Ward Blanton

Our Broad Present: Time and Contemporary Culture, Hans Ulrich Gumbrecht

Wrestling with the Angel: Experiments in Symbolic Life, Tracy McNulty

Cloud of the Impossible: Negative Theology and Planetary Entanglements, Catherine Keller

What Does Europe Want? The Union and Its Discontents, Slavoj Žižek and Srećko Horvat

Harmattan: A Philosophical Fiction, Michael Jackson

Nietzsche Versus Paul, Abed Azzam

Christo-Fiction: The Ruins of Athens and Jerusalem, François Laruelle

Paul's Summons to Messianic Life: Political Theology and the Coming Awakening, L. L. Welborn

Reimagining the Sacred: Richard Kearney Debates God with James Wood, Catherine Keller, Charles Taylor, Julia Kristeva, Gianni Vattimo, Simon Critchley, Jean-Luc Marion, John Caputo, David Tracy, Jens Zimmermann, and Merold Westphal, edited by Richard Kearney and Jens Zimmermann

A Hedonist Manifesto: The Power to Exist, Michel Onfray

An Insurrectionist Manifesto: Four New Gospels for a Radical Politics, Ward Blanton, Clayton Crockett, Jeffrey W. Robbins, and Noëlle Vahanian

The Intimate Universal: The Hidden Porosity Among Religion, Art, Philosophy, and Politics, William Desmond

Heidegger: His Life and His Philosophy, Alain Badiou and Barbara Cassin, translated by Susan Spitzer

The Work of Art: Rethinking the Elementary Forms of Religious Life, Michael Jackson

Sociophobia: Political Change in the Digital Utopia, César Rendueles, translated by Heather Cleary

There's No Such Thing as a Sexual Relationship: Two Lessons on Lacan, Alain Badiou and Barbara Cassin, translated by Susan Spitzer and Kenneth Reinhard